STUDENT RESOURCE GUIDE

DALE BUSKE
St. Cloud State University

EXCURSIONS IN MODERN MATHEMATICS

SEVENTH EDITION

Peter Tannenbaum
California State Univeristy – Fresno

Prentice Hall
is an imprint of

Reproduced by Pearson Prentice Hall from electronic files supplied by the author.

Publishing as Pearson Prentice Hall, Upper Saddle River, NJ 07458.

Printed in the United States of America.

ISBN-13: 978-0-321-57519-7
ISBN-10: 0-321-57519-9

1 2 3 4 5 6 BRR 12 11 10 09

Prentice Hall
is an imprint of

www.pearsonhighered.com

Table of Contents

Problem-Solving Strategies

When confronted with a mathematical problem, especially a mathematical word problem, the use of a strategy is fundamental. Effective problem-solving requires patience and appropriate technique. George Polya[1], whom many consider as one of the best problem solvers of all time, laid these four principles for problem solving.

1. Understand the problem.

2. Devise a plan or strategy.

3. Carry out the plan.

4. Look back.

Many strategies have been found to be effective in this framework. A few of these are listed below and can be found in the hints.

- ➡ Draw a diagram.
- ➡ Use a table or chart.
- ➡ Create an organized list.
- ➡ Solve a simpler (but similar) case first.
- ➡ Look for a pattern.
- ➡ Work backwards.
- ➡ Create a physical or graphical model.
- ➡ Make an educated guess and check.
- ➡ Use a variable.
- ➡ Look for symmetry.
- ➡ Use logical or deductive thinking.

[1] More information on effective problem-solving can be found in G. Polya, "How to Solve It", 2nd ed., Princeton University Press, 1957, ISBN 0-691-08097-6.

Chapter 1

Learning Outcomes

A successful student will be able to

- interpret and construct a preference schedule for an election involving preference ballots.
- accurately implement the plurality, Borda count, plurality-with-elimination, and pairwise comparisons vote counting methods.
- rank candidates in a preference election using recursive and extended methods.
- identify fairness criteria as they pertain to preferential voting methods.
- understand the significance of Arrows' impossibility theorem.

Skills to Help Prepare You for the Next Exam

At a minimum, be able to

- interpret and construct preference schedules for elections involving preference ballots (Exercises 1-10).
- determine the winner of a election using preference ballots using the
  - plurality method (Exercises 11, 12).
  - Borda count method (Exercises 17-22).
  - plurality-with-elimination method (Exercises 27-34).
  - method of pairwise comparisons (Exercises 35-38).
- rank candidates using
  - extended methods (Exercises 41-44).
  - recursive methods (Exercises 45-50).
- state the fairness criteria and identify when they are violated (Exercises 17-20, 27, 33-35, 59, 60).
- count the number of pairwise comparisons in a given election (Exercises 55, 56), the number of votes needed to win an election (Exercises 13-16), and Borda points given out (Exercises 23-26).
- state Arrows' impossibility theorem in your own words.

Study Tip

When it comes to implementing a particular algorithm, practice makes perfect. Practice will also make you more efficient at solving routine problems on exam day when time is short and your mind has other things to worry about.

Chapter 2

Learning Outcomes

A successful student will be able to

- represent a weighted voting system using a mathematical model.
- calculate the Banzhaf and Shapley-Shubik power distribution in a weighted voting system.

Skills to Help Prepare You for the Next Exam

At a minimum, be able to

- effectively use weighted voting terminology (Exercises 1, 2, 7, 8).
- construct a mathematical model of a weighted voting system (Exercises 3-6, 50, 51, 66).
- identify the presence of dictators, veto power, and dummies in a weighted voting system (Exercises 7-10, 52-56)
- compute Banzhaf and Shapley-Shubik power indices
  - by listing all (sequential) coalitions and identifying critical and pivotal players (Exercises 11-18, 25-28).
  - using techniques other than enumeration of all (sequential) coalitions (Exercises 19-24, 29-34, 45-50).
- compute factorials
  - using a calculator (Exercises 35, 36).
  - by hand (Exercises 37-40).
- determine possible values of the quota q for a weighted voting system (Exercises 3-6, 55, 60).
- count the number of coalitions having a particular property (Exercises 21, 22, 41-46)

Study Tip

It is often the case that a problem can be solved using more than one technique. For example, in theory one can always compute the Banzhaf and Shapley-Shubik power indices for a given weighted voting system by listing all critical and pivotal players. However, when the system has many voters, this process is (very!) long and tedious. Be on the lookout for special structure (e.g. equal number of votes, dictators, dummies, etc.) that will allow for other techniques which reduce the number of calculations needed.

Chapter 3

Learning Outcomes

A successful student will be able to

- state the fair-division problem and identify assumptions used in developing solution methods.
- recognize the differences between continuous and discrete fair-division problems.
- apply the divider-chooser, lone-divider, lone-chooser, and last-diminisher methods to continuous fair-division problems.
- apply the method of sealed bids and the method of markers to discrete fair-division problems.

Skills to Help Prepare You for the Next Exam

At a minimum, be able to

- quantify players' value systems (Exercises 1-6, 12-18, 29-42, 49-52, 57, 58, 67-70).
- identify fair shares to a given player (Exercises 7-11, 53-56).
- apply the following methods to solve continuous fair-division problems:
  - divider-chooser (Exercises 15-20, 76)
  - lone-divider (Exercises 21-32)
  - lone-chooser (Exercises 33-42)
  - last-diminisher (Exercises 43-52, 71, 72)
- apply the following methods to solve discrete fair-division problems:
  - sealed bids (Exercises 53-60).
  - method of markers (Exercises 61-70, 73, 74).
- add, subtract, multiply and divide fractions.

Study Tip

Carefully reading a mathematics textbook is an art that you should master. Read slowly--every word counts! Understand what you read before forging ahead. Mathematical terms and symbols have a very specific meaning—be sure that you understand the meaning of all of them. Very little information is repeated so you will likely want to read the text more than once (before and after class!). Working through the examples on paper as you read can keep you from nodding off and will add insight into what you read.

Chapter 4

Learning Outcomes

A successful student will be able to

- state the basic apportionment problem.
- implement the methods of Hamilton, Jefferson, Adams, and Webster to solve apportionment problems.
- state the quota rule and determine when it is satisfied.
- identify paradoxes when they occur.
- understand the significance of Balinski and Young's impossibility theorem.

Skills to Help Prepare You for the Next Exam

At a minimum, be able to

- compute standard divisors and quotas for a given apportionment problem (Exercises 1-10, 57).
- apply Hamilton's method (Exercises 11-18).
- apply Jefferson's and Adams's methods (Exercises 23-30, 33-40).
- apply Webster's method (Exercises 43-50).
- identify a paradox when it presents itself (Exercises 19-22, 55, 56).
- identify when the quota rule is violated (Exercises 31, 32, 41, 42, 55, 56).
- state Balinski and Young's impossibility theorem in your own words.

Study Tip

Taking time to write neatly and organize your thoughts can make all the difference in the world. This is especially true when the computations in a particular problem are straightforward. It often pays to slow down and carefully construct a table in order to avoid careless errors.

Chapter 5

Learning Outcomes

A successful student will be able to

- identify and model Euler circuit and Euler path problems.
- understand the meaning of basic graph terminology.
- classify which graphs have Euler circuits or paths using Euler's circuit theorems.
- implement Fleury's algorithm to find an Euler circuit or path when it exists.
- eulerize and semi-eulerize graphs when necessary.
- recognize an optimal eulerization (semi-eulerization) of a graph.

Skills to Help Prepare You for the Next Exam

At a minimum, be able to

- demonstrate an understanding of basic graph terminology (Exercises 1-4, 7-10, 15, 16, 45, 53).
- count a number of paths or circuits in a given graph (Exercises 11-14).
- draw a graph modeling a particular situation (Exercises 5, 6, 17-22, 47, 48).
- determine if a given graph has an Euler circuit or an Euler path (Exercises 23-28).
- find an Euler circuit or path for a graph having one (Exercises 29-36).
- find an (optimal) eulerization for a graph that has no Euler circuit (Exercises 37-40, 51, 52, 62).
- find an (optimal) semi-eulerization for a graph that has no Euler path (Exercises 41, 42, 47, 48, 58, 59).

Study Tip

A picture is worth a thousand words. A well-labeled diagram can lead to understanding many problems and can help solve many others. Try drawing diagrams even for problems that don't seem to lend themselves to this approach.

Chapter 6

Learning Outcomes

A successful student will be able to

- identify and model Hamilton circuit and Hamilton path problems.
- recognize complete graphs and state the number of Hamilton circuits that they have.
- identify traveling-salesman problems and the difficulties faced in solving them.
- implement brute-force, nearest-neighbor, repeated nearest-neighbor, and cheapest-link algorithms to find approximate solutions to traveling-salesman problems.
- recognize the difference between efficient and inefficient algorithms.
- recognize the difference between optimal and approximate algorithms.

Skills to Help Prepare You for the Next Exam

At a minimum, be able to

- find one or more Hamilton circuits or Hamilton paths for a given graph (Exercises 1-8, 57-63).
- calculate factorials by hand and with a calculator (Exercises 17-22).
- calculate how many edges and Hamilton circuits there are in a given complete graph (Exercises 23-28).
- use the brute-force algorithm to find optimal tours to traveling-salesman problems (Exercises 29-30, 53).
- find the nearest-neighbor tours and repetitive nearest-neighbor tours for traveling-salesman problems stated in various contexts (Exercises 29-36, 37-42, 69).
- find the cheapest-link tours for a traveling-salesman problems (Exercises 43-48, 70).
- determine the relative error of an approximate solution to a TSP (Exercises 49-52).

Study Tip

While calculators can handle many routine calculations, the ability to calculate by hand should not be dismissed. A great deal of understanding of mathematical concepts comes from cultivating this ability.

Chapter 7

Learning Outcomes

A successful student will be able to

- identify and use a graph to model minimum network problems.
- classify those graphs that are trees.
- implement Kruskal's algorithm to find a minimal spanning tree.
- understand Torricelli's construction for finding a Steiner point.
- recognize when the shortest network connecting three points uses a Steiner point.
- understand basic properties of the shortest network connecting a set of (more than three) points.

Skills to Help Prepare You for the Next Exam

At a minimum, be able to

- determine whether a graph is a tree or not (Exercises 1-8).
- find one or more (or all) of the spanning trees of a given network (Exercises 11-18, 51).
- calculate the redundancy of a given network (Exercises 11-14)
- find a minimal spanning tree for a weighted network (Exercises 19-24, 58).
- determine the shortest network connecting three points (Exercises 27-34).
- compute information about graphs containing 30-60-90° triangles (Exercises 43-50, 62, 63).
- reproduce Torricelli's construction (with a straightedge and compass).

Study Tip

Whenever an algorithm (or fact, construction, theorem, etc.) is named after a person, it is almost always important enough to master. Practice it enough to feel comfortable applying it in an exam environment in which you don't have the aid of a textbook or notes.

Chapter 8

Learning Outcomes

A successful student will be able to

- ➧ understand and use digraph terminology.
- ➧ schedule a project on *N* processors using the priority-list model.
- ➧ apply the backflow algorithm to find the critical path of a project.
- ➧ implement the decreasing-time and critical-path algorithms.
- ➧ recognize optimal schedules and the difficulties faced in finding them.

Skills to Help Prepare You for the Next Exam

At a minimum, be able to

- ➧ find indegrees, outdegrees, vertex-sets, arc-sets, cycles, etc. for a given digraph (Exercises 1-10).
- ➧ use digraphs to model real-life situations (Exercises 13-22).
- ➧ use a given priority list to schedule a project on *N* processors (Exercises 27-32, 35, 36, 69-71).
- ➧ use the decreasing-time algorithm to schedule a project on *N* processors (Exercises 37-44).
- ➧ apply the backflow and critical-path algorithms (Exercises 45-50).
- ➧ schedule independent tasks and determine how close the result is to optimal (Exercises 51-58).

Study Tip

Never underestimate the power of common sense. It is always wise to check your answer to a problem to see if it makes sense. It is doesn't, sometimes it is best to simply start over from the very beginning.

Chapter 9

Learning Outcomes

A successful student will be able to

- generate the Fibonacci sequence and identify some of its properties.
- identify relationships between the Fibonacci sequence and the golden ratio.
- understand the concept of gnomon and its relationship with the concept of similarity.
- recognize gnomonic growth in nature.

Skills to Help Prepare You for the Next Exam

At a minimum, be able to

- use and understand subscript notation (Exercises 1-6, 13, 14, 27, 28, 51).
- state and apply the defining characteristics of the Fibonacci numbers (Exercises 7,8, 25, 26, 52-54).
- state and apply basic properties of the golden ratio $\left(e.g.\ \phi^2 = \phi + 1\right)$ (Exercises 23, 24, 55, 57, 58, 68).
- evaluate expressions in which order of operations is essential (Exercises 19-22).
- work with scientific notation (Exercises 25, 26).
- solve quadratic equations (Exercises 29-32).
- determine when two figures are similar (Exercises 35, 36, 39, 40).
- determine parameters of gnomons to given plane figures (Exercises 41-50)
- state the *exact* value of the golden ratio $\left(\phi = \dfrac{1+\sqrt{5}}{2}\right)$ and an approximate value (1.618).

Study Tip

There is great beauty in mathematics. It is very common that patterns emerge in solving mathematical problems. Keep an eye open for these patterns and take advantage of them in solving problems when the opportunity arises.

Chapter 10

Learning Outcomes

A successful student will be able to

- understand percentages and percentage increases and decreases.
- differentiate between simple and compound interest and between annuities and installment loans.
- apply the compound interest formula to answer appropriate financial questions.
- understand what annual percentage yield represents.
- model exponential growth using geometric sequences.
- apply variations of the geometric sum formula (e.g. the fixed deferred annuity formula and the amortization formula) to understand deferred annuities and installment loans.

Skills to Help Prepare You for the Next Exam

At a minimum, be able to

- convert percentages to decimals and fractions and vice-versa (Exercises 1-4).
- solve word problems involving percentages (Exercises 5-20, 81-83).
- compute simple interest on an investment (Exercises 21-24).
- compute compound interest on an investment (Exercises 31, 32).
- compute the APY of an investment (Exercises 37-40, 43-46, 86).
- compute the original amount of principal for a given investment (Exercises 25, 26, 49, 50, 84, 85).
- determine explicit formulas for geometric sequences (Exercises 51-58).
- compute the sum of terms in a geometric sequence (Exercises 57-62).
- determine the future and present value of deferred annuities (Exercises 63-72).
- determine the future and present value of installment loans (Exercises 73-80).

Study Tip

Before memorizing any formulas, it is always important to understand why they work. This knowledge will make remembering the formula when it is needed much easier.

Chapter 11

Learning Outcomes

A successful student will be able to

- describe the four basic rigid motions of the plane and state their properties.
- classify the possible symmetries of any finite two-dimensional shape or object.
- classify the possible symmetries of a border pattern.

Skills to Help Prepare You for the Next Exam

At a minimum, be able to

- find the image of a shape under a reflection given the
  - axis of reflection (Exercises 1-4).
  - location of the image of one point (Exercises 4-8).
  - location of two fixed points of the reflection (Exercises 9, 10).
- find the image of a shape under a rotation given the
  - location of the rotocenter and the angle of rotation (Exercises 11, 12).
  - location of the image of two points (Exercises 15-18).
  - image of one point and the angle of rotation (Exercises 19, 20).
- find the image of a shape under a translation given the
  - translation vector (Exercises 21, 22).
  - image of a point under the translation (Exercises 23-26).
- find the image of a shape under a glide reflection given
  - a description of the translation and reflection (Exercises 27, 28).
  - the location of the image of two points (Exercises 29-34).
- give the symmetry type (D_N or Z_N) of finite figures in the plane (Exercises 39-44).
- give the symmetry type (mm, mg, m1, 1m, 1g, 12, 11) of border patterns (Exercises 49-52).
- identify a rigid motion based on its properties (Exercises 55, 56, 59-66, 68, 71).
- perform modular arithmetic (Exercises 13, 14).

Study Tip

Don't overlook the role mnemonic devices play in memorization. For example, in classifying border patterns, the second symbol is selected from the following (ordered) list: mg21. Your friends might give you "**m**ore **g**rief when you turn **21**." There are even folks that drink "**M**iller **G**enuine" but, of course, not until they are **21**.

Chapter 12

Learning Outcomes

A successful student will be able to

- explain the process by which fractals such as the Koch snowflake and the Sierpinski Gasket are constructed.
- recognize self-similarity (or symmetry of scale) and its relevance.
- describe how random processes can create fractals such as the Sierpinski Gasket.
- explain the process by which the Mandelbrot set is constructed.

Skills to Help Prepare You for the Next Exam

At a minimum, be able to

- calculate the perimeter and area of the figure found at any step in the construction of a given fractal (Exercises 1-14, 17-20, 23-30).
- draw the outcome of a particular step in the construction of a fractal (Exercises 13, 23).
- play the chaos game (Exercises 33-40).
- add and multiply complex numbers (Exercises 41-46).
- construct the Mandelbrot sequence for a given seed *s* and identify it as escaping, periodic, or attracted (Exercises 47-51, 62-65).

Study Tip

Spotting patterns is essential in mathematics. It takes patience. It is also often useful to *not* simply algebraic expressions and to resist the temptation to reach for a calculating device in the search for a pattern. It is also useful to use tables for format information in a manner that can easily be understood.

Chapter 13

Learning Outcomes

A successful student will be able to

- define key terminology in the data collection process.
- identify whether a given survey or poll is biased.
- list several sampling methods and discuss their advantages and disadvantages.
- identify components of a well-constructed clinical study.
- estimate the size of a population using the capture-recapture method.

Skills to Help Prepare You for the Next Exam

At a minimum, be able to

- identify the population, sample, and sampling frame (Exercises 1, 4, 6, 9, 11, 13, 17, 21, 41, 45, 46, 49, 50, 53, 54, 57).
- distinguish between a parameter and a statistic (Exercises 2, 3, 5, 13, 26, 65).
- identify the sampling method used in a survey or poll (Exercises 1, 6, 10, 13, 19, 23, 25, 27-30).
- identify whether sampling error is a result of sampling variability or bias (Exercises 3, 8, 11, 19, 23).
- estimate the size of a population using capture-recapture (Exercises 31-38).
- identify whether a study is blind, double-blind, or neither (Exercises 45, 47, 51, 55).
- list possible confounding variables in an experiment (Exercises 44, 61-63).
- discuss elements of poor design (Exercises 15, 20, 24, 43, 61, 62, 67, 69, 75).

Study Tip

Mathematics is not just a bunch of equations and formulas. Your instructor will expect you to communicate ideas in a precise and clear manner similar to what you may have encountered in an English course. When writing a paper, even in math class, always use complete sentences and proper grammar just as you would in any other class.

Chapter 14

Learning Outcomes

A successful student will be able to

- interpret and produce an effective graphical summary of a data set.
- identify various types of numerical variables.
- interpret and produce numerical summaries of data including percentiles and five-number summaries.
- describe the spread of a data set using range, interquartile range, and standard deviation.

Skills to Help Prepare You for the Next Exam

At a minimum, be able to

- read and produce frequency tables, bar graphs, pie charts, and histograms (Exercises 1-16).
- read and interpret histograms (Exercises 19-22).
- compute means and medians (Exercises 23-32, 65, 66).
- compute percentiles (including quartiles) (Exercises 33-40).
- compute five-number summaries and produce box plots (Exercises 41-44).
- interpret box plots (Exercises 45, 46).
- compute range, interquartile range, and determine the existence of outliers (Exercises 47-54).
- compute standard deviation (Exercises 55-58, 72).
- compute mode (Exercise set 59-64).

Study Tip

When you feel as if you have mastered a particular computational skill, put yourself in an exam environment and see how you fare. If all goes well, focus your attention on problems that appear to be stated a bit differently or are less computational (theoretical) in nature.

Chapter 15

Learning Outcomes

A successful student will be able to

- describe an appropriate sample space for a random experiment.
- apply the multiplication rule, permutations, and combinations to counting problems.
- understand the concept of a probability assignment.
- identify independent events and their properties.
- use the language of odds in describing probabilities of events.

Skills to Help Prepare You for the Next Exam

At a minimum, be able to

- describe an appropriate sample space for a given random experiment (Exercises 1-8).
- apply the multiplication rule to various counting problems (Exercises 9-18, 66, 67).
- compute the value of a given combination or permutation (Exercises 19-24, 31, 32).
- apply combinations or permutations to counting problems (Exercises 33-36, 63).
- construct a probability assignment for a given sample space (Exercises 37-42).
- describe an event using set notation (Exercises 43-48, 64, 65).
- find the probability of a given event when individual outcomes in a random experiment are equiprobable (Exercises 49-58, 70-77).
- write odds as probabilities and vice-versa (Exercises 59-62).

Study Tip

Probability is a topic that can produce answers that are not intuitive. It can also consist of problems that are very easy to state but difficult to solve. In calculating probabilities be sure to write out an appropriate sample space for the random experiment. Don't attempt to solve an entire probability (or mathematical!) problem in your head.

Chapter 16

Learning Outcomes

A successful student will be able to

- identify and describe an approximately normal distribution.
- state properties of a normal distribution.
- understand a data set in terms of standardized data values.
- state the 68-95-99.7 rule.
- apply the honest and dishonest-coin principles to understand the concept of a confidence interval.

Skills to Help Prepare You for the Next Exam

At a minimum, be able to

- find parameters (mean, median, standard deviation, etc.) of a normal distribution given in graphical form (Exercises 1-4).
- find the value of other parameters given two parameters of a normal distribution (Exercises 5-10).
- find standardized values given two parameters of a normal distribution (Exercises 15-18).
- use standardized values to determine parameters of a normal distribution (Exercises 21-28).
- determine the percent of data that falls between key values of a normal distribution (Exercises 29-40).
- apply the 68-95-99.7 rule to approximately normal distributions (Exercises 41-52).
- use the (dis)honest-coin principle to understand a random variable *X* (Exercises 57-66, 71).

Study Tip

A well-prepared student will not be surprised by exam questions. Solving as many different types of problems as possible can help you prepare much more than memorizing a particular example or equation. Mathematics is not a spectator sport. The best way to learn mathematics is to do mathematics.

Selected Hints

Chapter 1

WALKING

A. Ballots and Preference Schedules

1. (a) An example of a preference schedule can be found Table 1-1 in the text.

(b) To have a plurality is to have more than any other candidate.

3. (a) Eight voters ranked Professor Argand first on their ballot. Three voters ranked Professor Brandt first.

(b) In the U.S. Senate, a majority is 51 of the 100 Senators. In the U.S. House of Representatives, a majority is 218 of the 435 members.

(c) Go old-school. Use white-out on candidates *B*, *C*, and *E*.

5. (a) 20% is called a "threshold." How *many* votes does it equate to in this example?

(b) Use as few columns as you can in writing your schedule.

7. $(0.17)(1500) = 255$ voters prefer *L* the most; $(0.32)(500)$ voters prefer *C* the most. The remaining voters (of the 1500 total) prefer *M* the most.

9. Think of the numbers in each column as representing the position of the corresponding letter (candidate) from top to bottom in that column. So, according to the table, 47 voters rank the candidates as follows: *B*, *E*, *A*, *C*, *D*.

B. Plurality Method

11. (a) Under the plurality method, the candidate with the most first-place votes is declared the winner.

(b) *A*, for example, has 108+155 last-place votes. However, be careful to break the tie only by using candidates that have tied with the most first-place votes.

(c) Erase *A* and *C* from the preference schedule.

13. (a) Suppose that *A* receives *x* out of the remaining 30 votes and the competitor with the most votes receives the remaining $30 - x$ votes. What has to be true about the value of *x* to guarantee *A* at least a tie for first?

15. (a) Imagine that the votes are essentially split equally except that one candidate has an extra vote. That is, imagine dividing the votes equally among the candidates with any remaining votes (that cannot be divided equally) going to one candidate.

(b) The winning candidate has the smallest number of votes possible if the votes are distributed as evenly as possible among the candidates (with one candidate having a slight advantage).

C. Borda Count Method

17. (a) Under the Borda count method in this election, a candidate will receive 5 points for each first-place vote, 4 points for each second-place vote, 3 points for each third-place vote, 2 points for each fourth-place vote, and 1 point for each last-place vote.

(b) The preference schedule of the new election will contain four candidates (*A*, *B*, *C*, and *D*) since Epstein withdraws from the race. So, the voters will only list these 4 candidates on their preference ballots. The relative rankings on each ballot will remain the same.

(c) Professor *C* might feel a bit dumbfounded when Epstein is declared ineligible and the votes recounted.

(d) Review the fairness criterion stated in this chapter (i.e. majority, Condorcet, monotonicity, and IIA criterion).

19. Review what the *majority* criterion and *Condorcet* criterion state.

21. Strategy: Solve a simpler case. Imagine that there were 100 voters in this election. Determine how many voters correspond to each column before tallying Borda points.

23. (a) To achieve the maximum number of points possible, the election would need to be a landslide.

(b) If they received the minimum number of points possible, a candidate would be ranked really low on quite a few ballots wouldn't they?

(c) A voter gives out 4 points to the first-place candidate and fewer points to each lower ranked candidate.

(d) You would think it would be 110 times as much as the number of points given out by one voter.

(e) Here you need to assume there are 110 voters in the election. Of all the points possible (found in part (d)), *D* will get the number of points that remain.

25. Find how many total Borda points there are with 50 voters.

D. Plurality-with-Elimination Method

27. (a) Round 1:

Candidate	*A*	*B*	*C*	*D*	*E*
Number of first-place votes	8	3	5	5	0

E is eliminated. The number of first-place votes for the other candidates remains unchanged. Since no candidate has a majority of first-place votes, at least one more round of elimination is needed. In Round 2, candidate *B* is eliminated and their first-place votes transferred to *A*.

33. (b) A Condorcet candidate is one that beats every other candidate in head-to-head competition.

(c) Banzai Pipeline (Hawaii) would make a legitimate claim to supremacy. It might argue that Dungeons does not deserve to win.

E. Pairwise Comparisons Method

35. (a) With 5 candidates there are 4 + 3 + 2 + 1 = 10 pairwise comparisons to consider: *A* vs. *B*, *A* vs. *C*, *A* vs. *D*, *A* vs. *E*, *B* vs. *C*, *B* vs. *D*, etc. Give one point for each victory and ½ a point for each tie.

(b) With 4 candidates there are 3 + 2 + 1 = 6 pairwise comparisons to consider: *A* vs. *B*, *A* vs. *C*, *A* vs. *D*, *B* vs. *C*, *B* vs. *D*, and *C* vs. *D*.

39. With five candidates, there are a total of 10 pairwise comparisons (and hence 10 points distributed). Each candidate is part of four comparisons. Based on this, determine how many times each candidate will win.

F. Ranking Methods

41. (a) Rank the candidates according to how many first-place votes they each have.

(b) Rank the candidates according to how many Borda points they each have.

(c) Rank the candidates in reverse order of elimination.

(d) Rank the candidates according to how many points they receive from the comparisons.

45. From Example 1.5 we know that the winner using the Borda count method is *B* with 106 points. So, *B* is ranked first. Remove *B* from the preference schedule to form a new preference schedule. The winner of the election between candidates *A*, *C*, and *D* will be ranked second. Remove the winner of the election between *A*, *C*, and *D* to form a new preference schedule consisting of the two remaining candidates. The winner of that election will be ranked third and the loser fourth.

47. The winner, using plurality, is *A* with 10 first-place votes. So, *A* is ranked first. Remove *A* from the preference schedule to form a new preference schedule. The winner of the election between candidates *B*, *C*, and *D* using the plurality method will be ranked second. Remove the winner of the election between *B*, *C*, and *D* to form a new preference schedule consisting of the two remaining candidates. The winner of that election (using the plurality method of course) will be ranked third and the loser fourth.

49. From 41(d) we know that the winner using pairwise comparisons is *B*. Remove *B* from the preference schedule to form a new preference schedule among candidates *A*, *C*, and *D*. The winner of the election between candidates *A*, *C*, and *D* using pairwise comparisons will be ranked second. Repeat this process as needed.

G. Miscellaneous

51. Look at Example 1.14 again.

53. Strategy: Look for a pattern.
$501+502+503+\ldots+3220=$
$(1+2+3+\ldots+3220)-(1+2+3+\ldots+500)$

55. **(a)** Look at Example 1.15.

57. Strategy: Make an organized list.
Each column corresponds to a unique ordering of candidates *A*, *B*, and *C*.

JOGGING

61. Suppose the two candidates are *A* and *B* and that *A* gets *a* first-place votes and *B* gets *b* first-place votes and suppose that $a > b$. Then the preference schedule is

Number of voters	*a*	*b*
1st choice	*A*	*B*
2nd choice	*B*	*A*

Use this schedule to determine the winner using each of the four methods discussed in the chapter.

63. Strategy: Use logical or deductive thinking.
Suppose that in a reelection, the only changes in the ballots are in favor of the winning candidate. Could any other candidate pick up extra first-place votes?

67. **(a)** Strategy: Use a variable.
Suppose a candidate gets v_1 first-place votes, v_2 second-place votes, v_3 third-place votes, ..., v_N *N*th-place votes. Find the values of *p* (Borda point total) and *r* (Variation 2 total) and add them together.

(b) Suppose candidates C_1 and C_2 receive p_1 and p_2 points respectively using the Borda count as originally described in the chapter and r_1 and r_2 points under the variation described in this exercise. Show that if $p_1 < p_2$, we have $r_1 > r_2$.

69. **(a)** Strategy: Use a variable.
Each of the *x* voters gives out 25 Borda points to Ohio State. We find *x*.

(b) Strategy: Use a variable.
Let *x* = the number of second-place votes for Florida. Knowing the number of first-place votes and Borda count allows one to determine *x*. Use the fact that all of the voters listed these three teams in their top three.

71. Strategy: Use a variable.
By looking at Dwayne Wade's vote totals, it is clear that 1 point is awarded for each third-place vote. You are trying to determine *x*, the number of points awarded for each second-place vote.

73. **(a)** With 23 last-place votes, *A* is eliminated first.

(b) Strategy: Make an educated guess and check. Think of an example in which a candidate has a plurality of first-place votes, but many last-place votes leading to early elimination.

(c) Strategy: Make an educated guess and check. Try to eliminate a "middle" candidate by moving the winning candidate up on some ballots. Arrange the numbers so that a different candidate winds up winning.

Chapter 2

WALKING

A. Weighted Voting Systems

1. **(a)** Inside the square brackets we always list the quota first, followed by a colon and then the respective weights of the individual players.

(b) The total number of votes is the sum of the weights of the individual players.

(c) The weight of P_1 is the first number after the colon. The weight of P_2 is the second number listed after the colon.

(d) It takes 65 votes to pass a motion. What percentage of the total vote is this number?

3. **(a)** The quota *q* must be a whole number more than half of the total votes. Set up an inequality to determine q. After you finish, be sure to look back to be certain the quota you find works.

(b) The quota q should not be larger than the total number of votes.

(c) Strategy: Solve a simpler problem. See 3(a) for a simpler problem. In the inequality you used, simply use 2/3 rather than 1/2.

(d) Be sure to read the problem carefully. How is this question different than 3(c)?

5. (a) Strategy: Use a variable. The quota is just a bit more than half the number of total votes. The value of x must be an integer that makes half the total number of votes just smaller than 49.

7. A player is a dictator if the player's weight is bigger than or equal to the quota. A player that is not a dictator, but that can single-handedly prevent the rest of the players from passing a motion is said to have veto power. A dummy is a player with no power.

At most one player could be a dictator (P_1). Several players (perhaps even all of them) can have veto power. The most likely candidates for veto power are those with the most votes. [Can P_1 prevent a motion from passing? Can P_2 ? P_3 ? etc.] The most likely candidates for dummies are those with the fewest votes.

9. (a) All of the players would be required in order for a motion to pass if this were the case. How small could the value of q be?

(b) In order that P_2 have veto power, the other players should not be able to pass a motion without their votes. That is, the quota must be more than the number of votes the other players control.

B. Banzhaf Power

11. (b) Winning coalitions are those groups of players that have enough votes to pass a motion (i.e. meet the quota q). Here are all of the coalitions (winning and losing): $\{P_1\}$, $\{P_2\}$, $\{P_3\}$, $\{P_4\}$, $\{P_1, P_2\}$, $\{P_1, P_3\}$, $\{P_1, P_4\}$, $\{P_2, P_3\}$, $\{P_2, P_4\}$, $\{P_3, P_4\}$, $\{P_1, P_2, P_3\}$, $\{P_1, P_2, P_4\}$, $\{P_1, P_3, P_4\}$, $\{P_2, P_3, P_4\}$, $\{P_1, P_2, P_3, P_4\}$

(c) A critical player is one whose desertion turns a winning coalition into a losing coalition.

(d) For each winning coalition found in part (b), determine the number of times each player is critical. According to Banzhaf, a player's power is proportional to how often they are critical relative to the other players.

13. (a) The winning coalitions are $\{P_1, P_2\}$, $\{P_1, P_3\}$, and $\{P_1, P_2, P_3\}$. Underline the critical player(s) in each.

15. (b) The quota is one more than in (a), so some winning coalitions may now be losing coalitions. For the ones that are still winning, any players that were critical in (a) will still be critical, and there may be additional critical players. A quick check shows that $\{P_1, P_2, P_5\}$, $\{P_1, P_3, P_4\}$, and $\{P_2, P_3, P_4, P_5\}$ are now losing coalitions (they all have exactly 10 votes).

19. (a) Strategy: Use logical thinking. A player is critical in a coalition if that coalition without the player is not on the list of winning coalitions.

23. Strategy: Make an organized list. The key to understanding this problem is to understand how much power D has. The other players will (obviously) share the remaining power equally. The winning coalitions have two, three, or four players. List them. How often is D critical in each?

C. Shapley-Shubik Power

25. (a) Strategy: Make an organized list. There are 3! = 6 sequential coalitions of the three players. $\langle P_2, P_3, \underline{P_1} \rangle$ is one of them with the pivotal player underlined. See also Example 2.17.

(b) P_1 is underlined 4 times; P_2 is underlined 1 time; P_3 is underlined 1 time. Each underline adds to Shapley-Shubik power.

27. (a) Don't do any calculations.

(b) Strategy: Make a table or chart.
There are 4! = 24 sequential coalitions of the four players. The pivotal player in twelve of these sequential coalitions is underlined below.
$< P_1, \underline{P_2}, P_3, P_4 >$, $< P_1, \underline{P_2}, P_4, P_3 >$,
$< P_1, \underline{P_3}, P_2, P_4 >$, $< P_1, \underline{P_3}, P_4, P_2 >$,
$< P_1, P_4, \underline{P_2}, P_3 >$, $< P_1, P_4, \underline{P_3}, P_2 >$,
$< P_2, \underline{P_1}, P_3, P_4 >$, $< P_2, \underline{P_1}, P_4, P_3 >$,
$< P_2, P_3, \underline{P_1}, P_4 >$, $< P_2, P_3, P_4, \underline{P_1} >$,
$< P_2, P_4, \underline{P_1}, P_3 >$, $< P_2, P_4, P_3, \underline{P_1} >$.

(c) Focus on P_1 and P_2 only.

31. (a) A player is pivotal in a sequential coalition if the players listed prior to that player form a losing coalition while that player *together* with the prior players form a winning coalition.

33. Strategy: Use logical thinking.
D will clearly never be pivotal in the first position. *D* is also never pivotal is the second position since it loses all tie votes. How about when *D* is in the third or fourth positions?

D. Miscellaneous

35. (a) Look for the *x*! or *n*! key. On some calculators, you may need to look under a probability (PRB) menu.

(d) Pay attention to the appropriate units in your calculation. They will help guide you to the solution.

37. (a) Strategy: Look for a pattern.
$10! = 10 \times (9 \times 8 \times \ldots \times 3 \times 2 \times 1)$

(c) Strategy: Look for a pattern.
$11! = 11 \times 10 \times (9 \times \ldots \times 3 \times 2 \times 1)$

39. (a) Strategy: Look for a pattern.
$$\frac{9!+11!}{10!} = \frac{9!}{10!} + \frac{11!}{10!} = \frac{9!}{10 \times 9!} + \frac{11 \times 10!}{10!}$$

43. (b) Look back to 41(b), but use sequential coalitions rather than coalitions in the calculation.

45. (c) The remaining players divide the remaining power equally.

JOGGING

47. If x is even, then $q = \frac{15x}{2} + 1$. If x is odd, then $q = \frac{15x+1}{2}$. In either case, your goal is to show that $8x \geq q$.

49. Since P_2 is a pivotal player in the sequential coalition $< P_1, P_2, P_3 >$, it is clear that $\{P_1, P_2\}$ and $\{P_1, P_2, P_3\}$ are winning coalitions, but $\{P_1\}$ is a losing coalition. Each other sequential coalition gives similar information.

53. (a) What would happen if a winning coalition that contains *P* (a dummy) is not a winning coalition without *P*?

(b) Link being a dummy to never being critical.

55. (a) There are rules regarding the quota. See Examples 2.2 and 2.3.

(b) Only one player P_1 could possibly be a dictator. They have at least as many votes as the quota *q*.

(c) To have veto power is to have less power than a dictator. If only one player can have veto power, it must be that player with the largest number of votes.

(d) That is, which values of *q* result in P_1 and P_2 or all three players having veto power?

57. (a) Strategy: Look for a pattern.
[24: 14, 8, 6, 4] is just [12: 7, 4, 3, 2] with each value multiplied by 2. It is like describing the system using pints rather than quarts. To fill a bucket with water (pass a motion), it takes the same amount of water in both cases (whether it is measured in quarts or pints). If each player has the same amount of water in each system, the amount of power they each hold in each is the same.

59. There are two key cases to consider: either the senior partner is the first player listed in a sequential coalition or not. The senior partner will always be critical when not listed first. Once the senior partner's power index is known, the *N* junior partners will share the remaining power equally.

61. Strategy: Use a table or chart.
Consider your power in the systems [8: 5, 4, 2, 2], [8: 6, 3, 2, 2], and [8: 6, 4, 1, 2]. An applet may save some time in calculation.

65. **(a)** Strategy: Use a table or chart.
List all of the possibilities in a table or chart. For example, the complement of the losing coalition $\{P_1\}$ is the winning coalition $\{P_2, P_3\}$.

(b) Strategy: Use a table or chart.
List all of the possibilities in a table or chart. One losing coalition (with 2 votes) is $\{P_2, P_3\}$. The complement of this coalition, $\{P_1, P_4\}$, is a winning coalition (with 3 votes).

(c) Strategy: Use logical thinking.
If P is a dictator, what can be said about any winning coalition? That is, which player is certain to be a part of every winning coalition?

(d) In a decisive voting system, each losing coalition pairs up with a winning coalition (its complement). See, for example, (a) and (b).

67. **(a)** In each nine-member winning coalition, every member is critical. What happens in coalitions consisting of 10 or more members?

(b) At least nine members are needed to form a winning coalition. So, there are 210 + 638 = 848 winning coalitions. Since every member is critical in each nine-member coalition, the nine-member coalitions yield a total of $210 \times 9 = 1890$ critical players. How many critical players are there in the coalitions consisting of 10 or more members?

(c) Basically, you are being asked to explain why each permanent member is critical in each of the 848 winning coalitions.

(d) The five permanent members together have $5 \times 848/5080$ of the power. The remaining power is shared equally.

Chapter 3

WALKING

A. Fair Division Concepts

1. **(a) & (b)** Strategy: Use a variable.
Let C = the value of the chocolate half (in Angelina's eyes). Let S = the value of the strawberry half (in Angelina's eyes). Then $C = 3S$ and $C + S = \$24$.

(c) The piece shown is a particular fraction of the strawberry half.

3. **(a)** Strategy: Use a variable.
Let M = the value of the mushroom half (in Homer's eyes). Let P = the value of the pepperoni half (in Homer's eyes). What can be said about M and P?

(c) Once you know what the mushroom half is worth to Homer, you can use a proportion.

5. **(a)** Strategy: Use a variable.
Let C = the value of the chocolate part (in Karla's eyes). Let S = the value of the strawberry part (in Karla's eyes). Let V = the value of the vanilla part (in Karla's eyes). Then, $S = 2V$, $C = 3V$, and $C + S + V = \$30$.

(b) With six players, a fair share is worth at least 1/6 the value of the entire cake.

7. **(a)** A fair share is a slice worth 1/3 (in this case since there are three players) of what *that* player values the entire cake. How much does Ana value the entire cake?

(d) There is only one piece that Ben considers a fair share. He must get that piece in any fair division.

9. **(a)** Strategy: Make a table or chart.
Completing the table with one more column might be helpful.

(d) Which player would accept s_1?

11. **(a)** What is a fair share worth to Abe?

(b) What is a fair share worth to Betty?

(e) To start, Dana must receive s_1.

13. **(a)** Strategy: Use a variable.
Let x denote the value of slices s_2 and s_3 to Adams. Then, determine the value of the expression
$(x+\$40,000)+x+x+(x+\$60,000)$.

(b) Strategy: Use a variable.
Let x denote the value of slice s_3 to Benson. Then,
$x+(x+\$8000)=0.40(\$400,000)$.

(e) To start, Duncan must receive s_4. This assignment also determines Adams' fair share.

B. The Divider–Chooser Method

15. **(a)** Strategy: Solve a simpler problem first.
IF Jared had to cut the sandwich into FOUR pieces each worth the same amount, how might he cut it? Use the fact that three little meatball sections are each worth the same as that big vegetarian half.

(b) Karla wants a half with lots of vegetarian and little meatball if she can get it.

17. **(a)** Strategy: Use a variable.
Let x denote the amount Martha values each inch of turkey (or roast beef). Then, decide how she values each inch of the entire sandwich.

(b) Strategy: Use a variable.
As in (a), let y denote the amount that Nick values each inch of ham (and turkey). Use this to determine how he values each inch of the sandwich. This will determine how he will choose and the value of his choice.

19. **(a)** If David is rational, he will hedge his bets so that no matter which piece Paula chooses, he ends up with at least half the value of the pizza.

C. The Lone-Divider Method

21. **(a)** Strategy: Make an organized list.
Since neither chooser bid on s_1, that piece must go to D.

(b) Strategy: Make an organized list.
Consider two cases: (1) C_1 receives s_2 and (2) C_1 receives s_3.

23. **(a)** Strategy: Use logical thinking.
C_1 must receive s_2. From there, C_2's fair share is determined.

(b) Consider two cases: (1) C_1 receives s_2 and (2) C_1 receives s_3.

(c) C_1 must receive s_2. From there, consider two cases: (1) C_2 receives s_1 and (2) C_2 receives s_3.

25. **(a)** Obviously the three choosers do not believe s_3 is worth ¼ of the value of the original plot of land. So would letting it go be such an issue for them?

(b) Strategy: Solve a simpler problem first.
Think about a situation in which s_1 is worth 90% (or more) of the value of the plot of land to C_2 and C_3.

(d) Strategy: Solve a simpler problem first.
Think about a situation in which s_2 is worth 51% of the value of the plot of land to C_2 and C_3.

27. **(a)** If C_1 is to receive s_2, then C_2 receives s_4 (and conversely). What does this force C_4 to receive?

29. **(a)** Strategy: Use logical thinking.
The Divider will be the only player that could possibly value each piece equally.

(b) Strategy: Make a table or chart.
To determine the bids placed, each player's bids should add up to \$480,000.

(c) Use the player's bid lists that were found in part (b).

31. **(a)** Jared likes everything equally and he wants to cut the sandwich into three parts all of equal value. Seems pretty simple, doesn't it?

(b) Imagine that Karla values each two inch segment of vegetarian sub equally (she does). Ignore the meatball part.

(c) Strategy: Use a variable.
Find the value (as a percentage?) of each inch of the sandwich to Lori. Call that value x.

D. The Lone-Chooser Method

33. (a) Angela sees s_1 as being worth \$18. In her second division, she will create three \$6 pieces.

(b) Strategy: Use a variable.
Boris sees the right half of the cake as being worth \$15. So, in his second division, he will create three \$5 pieces. Suppose he cuts a x° wedge out of the strawberry part. Then,

$$\frac{x^\circ}{90^\circ} \times \$9.00 = \$5.00$$

(c) Since Carlos values vanilla twice as much as strawberry, Carlos would want to take the most vanilla that he could from Angela and from Boris.

35. Boris chooses s_2 because he views it as being worth \$12 + \$6 = \$18. He will cut s_2 into portions that are each worth \$6.

37. After Arthur makes the first cut, Brian chooses a piece worth (to him) 100% of the cake! Awesome, dude.

(a) Basically, s_1 is just like one having all the same flavor (in terms of value to him).

(b) Arthur places all of the value on the orange half of s_2.

(c) Carl is going to choose as much chocolate and vanilla as he can get his hands on.

(d) Brian knows he will do well (better than 1/3 in the end) right after Arthur cuts the cake. Arthur, on the other hand, has minimized his own pleasure with his cut. Carl is the real winner gaining a piece that is worth far more than 1/3 of the cake in the end.

41. (a) Remember, Karla is a vegetarian so she will simply divide the vegetarian part.

(b) Based on his tastes, Jared will always divide any piece into equally sized portions.

(c) Karla will once again divide whatever vegetarian part she has into equally sized portions.

(d) When she has a choice, Lori prefers meatball subs to equally sized vegetarian subs. She values the vegetarian half of the sandwich as 1/3 and the meatball half of the sandwich as 2/3.

E. The Last-Diminisher Method

43. (a) Any player that values the piece when it is their turn as worth more than \$6.00 will become a diminisher.

(b) Remember, if a player diminishes the piece in round 1, they make it worth exactly 1/5 of the value of the entire cake.

(c) As bidders, those that were not the last diminisher in round 1 viewed the piece that was taken as being worth less than \$30/5 = \$6.

47. (a) The method is called "last-diminisher" for a reason.

(b) Imagine 12 siblings sitting around a table dividing the land using this method. Suppose that the order of play is "clockwise."

49. (a) Since Boris values strawberry more than Carlos, he will diminish in round 1. So too will Angela (by 0%). The key is figure out how much of the strawberry wedge (what fraction) Boris will diminish. To this end, a proportion will work.

(c) First determine what the cake is worth to Carlos at the beginning of round 2. Cut a piece of vanilla that has half of that value to him. Again, use a proportion.

(d) Determine the value, in dollars, each player receives. The fraction will be that value over what they value the whole cake to be.

51. (a) Determine the fraction of the whole that an inch of meatball sub is worth to Lori. Use it to find the fraction of the whole that an inch of the vegetarian sub is worth to her. This value times x should then be set equal to 1/3 (to determine the marker for Lori's fair share).

(b) Karla will diminish the C-piece cut by Lori because the vegetarian part is even more valuable to her.

(c) With two players remaining, the process is essentially divider-chooser.

F. The Method of Sealed Bids

53. (a) Strategy: Make a table or chart.
Ana's bids total \$900 so that a fair share to her is 1/3 of that amount. Remember, each player may have a different fair share. Organize your calculations in a table or chart.

(b) Since the value of the item she received (the desk) is \$180, Ana's preliminary cash settlement is \$120 (these two values yield her fair share).

(c) The surplus is the difference between the amount paid in and the amount paid out in the first settlement. The fact that a surplus exists is what makes the method of sealed bids so interesting (and appealing).

(d) Split the surplus three ways.

57. Strategy: Use a variable.
Suppose that Angelina bid \$*x* on the laptop. Then, she values the entire estate at *x*+\$2900. Brad, on the other hand, values the entire estate at \$4640. Complete the table below to determine the value of *x*.

	Angelina	Brad
Fair Share	(*x*+\$2900)/2	\$2320
Value of items received	*x*+\$300	\$2780
Prelim. cash		-\$460
Share of surplus		
Final cash	\$355	-\$355

G. The Method of Markers

61. The key is to spot the first first marker and the first second marker.

67. (a) Quintin thinks the total value is $3 \times \$12 + 6 \times \$7 + 6 \times \$4 + 3 \times \$6 = \$120$, so to him a fair share is worth \$30. He would place his markers after every \$30 worth of comic books.

(b) Look for the first first marker, first second marker, first third marker, and the last third marker.

JOGGING

71. (a) The total of the valuable area is $30{,}000\,m^2$. What is the area of *C* ? [Is it 1/3 of the valued area?]

(b) Strategy: Use a variable.
A cut parallel to Park Place which divides the parcel in half is illustrated below. Suppose that the cut is made *x* meters from the bottom. Use the fact that $\frac{y}{x} = \frac{60}{100}$ and that the area of the bottom trapezoid is be $11{,}000\,m^2$ to determine the value of *x*.

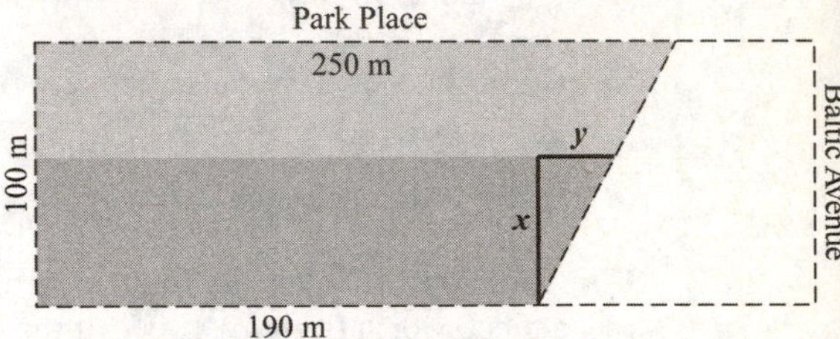

(c) Strategy: Use a variable.
Find *x* so that each shaded area is $11{,}000\,m^2$.

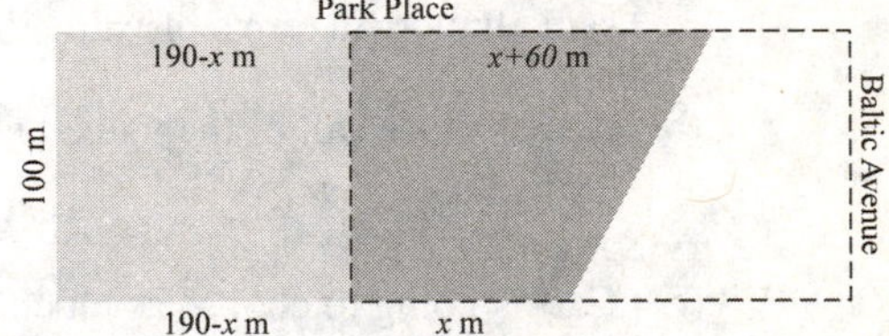

73. Strategy: Make a table or chart.
In the first settlement *A* receives \$*x*/2; *B* receives the partnership and pays \$*y*/2. Calculate the surplus and give half to each player to determine the final settlement.

75. Strategy: Make an educated guess and check.
Start by letting *B* get the strawberry part. Then, try giving *C* an all chocolate piece worth 50% (in their eyes) and see what remains for *A*.

79. When all the bids are negative, give the item (chore) to the least negative (greatest) bid.

Chapter 4

WALKING

A. Standard Divisors and Standard Quotas

1. **(a)** The standard divisor represents the number of people represented per seat.

(b) The standard quota of Apure is the population of Apure divided by the standard divisor. It represents, in a perfectly divisible world, the number of seats in Congress that Apure deserves.

3. **(a)** The "states" in any apportionment problem are the entities that will have "seats" assigned to them according to a share rule. The standard divisor is the number of "seats" that each "state" is accorded.

(b) That is, can you see the analogy between "states" and "seats" and "routes" and "buses?"

5. **(a)** If the seats were divisible, each state would receive their standard quota.

(c) The population of each state is their standard quota × the standard divisor.

7. With 7.43% of the U.S. population, Texas would ideally receive that percentage of the number of seats available in the House of Representatives.

B. Hamilton's Method

11. Find and sum the lower quotas. Then, allocate any additional seats according to those states whose standard quota have the largest fractional part.

19. **(c)** Notice that for studying an extra two minutes (an increase of 3.70%), Bob benefits. However, Peter, who studies an extra 12 minutes (an increase of 4.94%), loses.

21. **(c)** Jim, a new "state," enters the discussion and is given his fair share (six pieces) of candy.

C. Jefferson's Method

23. Strategy: Make an educated guess and check. You will need to find a modified divisor (smaller than the standard divisor) for this problem. Luckily, quite a few modified divisors will work. Choose a modified divisor so that the modified lower quotas (the modified quotas rounded down) sum to 160.

25. Strategy: Make an educated guess and check. Any modified divisor between approximately 971 and 975.7 can be used for this problem.

31. Jefferson's method only allows for certain types of quota violations.

D. Adams' Method

33. Strategy: Make an educated guess and check. You will need to find a modified divisor (larger than the standard divisor) for this problem. This modified divisor will need to be chosen so that the modified upper quotas sum to 160. The modified divisor you choose will need to be less than 51,000.

41. The difference between the number of seats that California would receive under Adams's method and California's standard quota is greater than 1. What fairness criteria does this violate?

E. Webster's Method

43. Strategy: Make an educated guess and check. Any modified divisor between approximately 49,907 and 50,188 can be used for this problem. Remember too that conventional rounding is the name of the game in Webster's method.

49. Any modified divisor between approximately 0.499% and 0.504% can be used for this problem.

JOGGING

53. What can be said about how big $\frac{P_A}{D}$ must be (here D is Jefferson's modified divisor)? And how small must $\frac{P_B}{D}$ be?

55. **(a)** The fractional parts of the quotas must add up to 1. Moreover, one is bigger than the other. So how does rounding work in such a case?

(b) Remember that Webster's method can never suffer from the Alabama or population paradox.

(c) Remember that Hamilton's method satisfies the quota rule.

57. (a) Think proportions.

(b) Use (a) and again think proportions.

59. (a) Any modified divisor between approximately 105.3 and 107.1 can be used for Adams' method.

(b) For $D = 100$, the modified quotas are just a bit too large. For $D < 100$, each of the modified quotas will increase (which isn't very helpful knowing that $D = 100$ didn't work). For $D > 100$, each of the modified quotas will decrease (in fact, they will decrease more than you might like).

(c) What other choices for modified divisors are there if $D < 100$, $D = 100$, and $D > 100$ do not work?

61. The fact that each state receives either its upper quota or lower quota is key in this exercise.

63. (a) Both the fractional part and the relative fractional part of the quota for one state should be smaller than for the other state.

(b) Try making the fractional part of the quota for *B* bigger than that of *A* and at the same time the relative fractional part of *A* bigger than that of *B*.

(c) Assume that $f_1 > f_2$, so under Hamilton's method the surplus seat goes to *A*. Using Lowndes' method, under what condition would the surplus seat go to *B*?

mini-excursion 1

A. The Geometric Mean

1. (a) The geometric mean is the average of two numbers under multiplication (rather than addition). The geometric mean of two positive numbers *a* and *b* is $\sqrt{a \cdot b}$.

(b) Each number is twice as large as in part (a). So how do you expect the average to behave? You can do this in your head. If in doubt, use the formula.

(c) $\sqrt{\frac{1}{N}} = \frac{1}{\sqrt{N}}$ for any positive number *N*.
So, with (b) you can do this in your head.

3. Strategy: Try a simpler problem first.
You might try to find the geometric mean of 2^3 and 2^5. [Note: It is $\sqrt{2^3 \times 2^5} = \sqrt{2^8} = 2^4$.] When you finish this problem, look back and see if you can make a general conclusion.

5. See Example ME1.1.

7. Strategy: Look for a pattern.
If you do the calculations correctly, you will spot a pattern in the third column.

9. (a) You will need to do a little algebra to simplify $\sqrt{(k \cdot a) \cdot (k \cdot b)}$.

11. (a) Recall the Pythagorean Theorem which says that $(AB)^2 + (BC)^2 = (AC)^2$.

(b) Remember that the hypotenuse is always the longest side of a right triangle.

B. The Huntington-Hill Method

13. The cutoff for state *A* is the geometric mean $\sqrt{25 \times 26}$. For state *B*, it is $\sqrt{18 \times 19}$.

17. (a) Under the Huntington-Hill method the standard divisor does not work but a modified divisor of $D = 990$ works!

(b) Aleta has a standard quota of 86.915 and yet an unusual number of representatives in the final apportionment. What is the name of this type of violation?

C. Miscellaneous

19. If $a < b$, then $\sqrt{a} < \sqrt{b}$. It follows that $\sqrt{c} - \sqrt{a} > \sqrt{c} - \sqrt{b}$. Square both sides, expand, and do a little algebra.

21. **(b)** Since $a < b,\ a^2 + a^2 < a^2 + b^2 < b^2 + b^2$. Divide each quantity in half.

(c) Note that

$$\sqrt{\left(\frac{b-a}{2}\right)^2 + \left(\frac{b+a}{2}\right)^2} \geq \sqrt{\left(\frac{b+a}{2}\right)^2}.$$

Chapter 5

WALKING

A. Graphs: Basic Concepts

1. **(a)** The basic elements of a graph are dots and lines (vertices and edges in mathematical language). The vertex set is a listing of the labels given to each dot (vertex).

(b) Remember, each edge can be described by listing the pair of vertices that are connected by that edge.

(c) The degree of a vertex is the number of edges meeting at that vertex.

5. Strategy: Draw a diagram.
There are an infinite number of ways to do this. For example, one way would be to place all of the vertices on a line in one picture and in the shape of a polygon in the other.

7. **(a)** Remember that two vertices are adjacent in a graph if they are joined by an edge.

(b) Two edges are adjacent if they share a common vertex.

(d) Each edge contributes two degrees to the total.

9. **(a)** Each vertex will have an edge "coming in and going out." Think symmetrically.

(b) Think "three pieces."

(c) Again, as an example, "think three pieces."

11. A path is a sequence of vertices with the property that each vertex in the sequence is adjacent to the next one. Remember that an edge can be part of a path only once. The number of edges in the path is called the length of the path.

The multiplication rule may come in handy for counting the number of paths between two vertices. For example, there are 3 paths from *H* to *D*. Why? There is exactly 1 path from *H* to *A* and there are three paths (*A*, *D*; *A*, *B*, *C*, *D*; and *A*, *B*, *D*) from *A* to *D*. Putting these facts together gives $1 \times 3 = 3$ paths from *H* to *D*.

Any path from *C* to *F* must pass through *A* and *H*. The total number of these is the number of paths from *C* to *A* times the number of paths from *A* to *H* times the number paths from *H* to *F*.

13. Strategy: Make an organized list.
There is only one circuit of length 1. Circuits of length 2 are hard to come by since they would only consist of two vertices and two edges between those vertices. To count the total number of circuits, note that there are not any circuits having length longer than 4 in this graph.

15. An edge whose removal makes a connected graph disconnected is called a bridge. Graphs can have no bridges, some bridges, or all bridges as its edges.

B. Graph Models

17. Strategy: Draw a diagram.
A graph model will use one vertex to represent North Kingsburg and one vertex to represent South Kingsburg (along with three other vertices – one for each island).

21. Strategy: Draw a diagram.
Let the vertices represent the teams (so there are six vertices). The edges will correspond to the tournament pairings.

C. Euler's Theorems

23. An Euler *circuit* is a circuit that passes through every edge of a graph. Euler's Circuit Theorem can be used to decide if a given graph has an Euler circuit.

Is the graph connected? Are the vertices all even?

An Euler *path* is a path that passes through every edge of a graph. Euler's Path Theorem can be used to decide if a given graph has an Euler path.

Is the graph connected? Are all but exactly two of the vertices even?

D. Finding Euler Circuits and Euler Paths

29. Fleury's algorithm will always find an Euler circuit for a graph that has one. Typically when a graph has one Euler circuit, it has several. So several answers are possible.

31. You have three choices at this point. One choice is really bad? Do you see why?

33. In order to have an Euler path, a graph must have exactly two odd vertices. Start by finding these – they will be the beginning and the end of an Euler path. Once again, Fleury's algorithm can typically produce several possible Euler paths.

F. Eulerizations and Semi-eulerizations

37. The key idea in eulerization is to turn the odd vertices into even vertices by adding "duplicate" (not new) edges in strategic places. Adding as few duplicate edges as possible gives an optimal eulerization.

A good place to start is to identify (color?) the odd vertices. Then, by connecting pairs of these with sequences of existing edges, the odd vertices will vanish. The difficult part will be adding the fewest number of edges as possible to accomplish this task.

The odd vertices and two possible duplicate edges have been identified for you below.

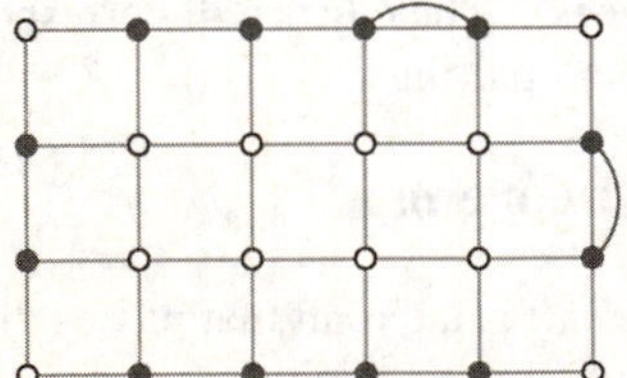

41. The key idea in semi-eulerization is to turn all *but two* odd vertices into even vertices by adding "duplicate" (not new) edges in strategic places. Adding as few duplicate edges as possible gives an optimal semi-eulerization.

G. Miscellaneous

47. Strategy: Draw a diagram.
Model the tennis court using a graph in which the vertices (at least those of odd degree) are located at "junction points." Then, find a semi-eulerization of that graph.

JOGGING

49. (a) Strategy: Solve a simpler problem first. If you were to add a new edge to the graph, how many vertices would increase in degree? Try an example using a small graph first.

(b) What would the sum of the degrees of all the vertices be?

51. Strategy: Use logical thinking.
What can be said about the edge incident to a vertex of degree 1?

53. (a) In order for each vertex to be even, each vertex in group A and group B must be even. Each vertex in group A connects to each vertex in group B. So this tells us information about the number of vertices in group B.

(b) See what happens if (i) $m = 1$, (ii) $m = 2$, and (iii) $m > 2$.

55. (a) Experiment by building graphs for $N = 4$, $N = 5$, and $N = 6$. The strategy that will give you the most money will become clear.

(b) See Example 1.14. Really.

Chapter 6

WALKING

A. Hamilton Circuits and Hamilton Paths

1. A Hamilton *circuit* in a graph is a circuit in a graph that visits each vertex of the graph once and only once. Similarly, a Hamilton *path* in a graph is a path in a graph that visits each vertex of the graph once and only once.

In this exercise, one Hamilton circuit is *A*, *D*, *C*, *E*, *B*, *G*, *F*, *A*.

3. Strategy: Make an organized list.
There are four Hamilton circuits in (a). Two are the mirror images of the other two. There

are four Hamilton circuits in (b). As in (a), two are mirror images of the other two. Eight Hamilton circuits can be found in (c).

Note that edges *CD* and *DE* must be a part of every Hamilton circuit (in (a), (b), and (c)) and that *CE* cannot be a part of any Hamilton circuit.

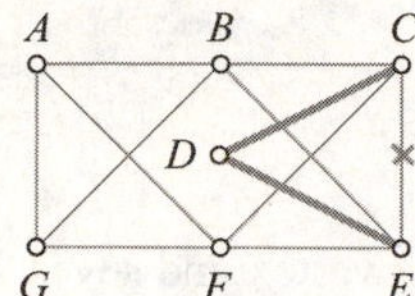

There are 2 possibilities at vertex *C*: edge *BC* or edge *CF*. Edge *BC* forces edge *FE* (otherwise there would be a circuit *E*, *B*, *C*, *D*, *E*) and edge *CF* forces edge *EB*.

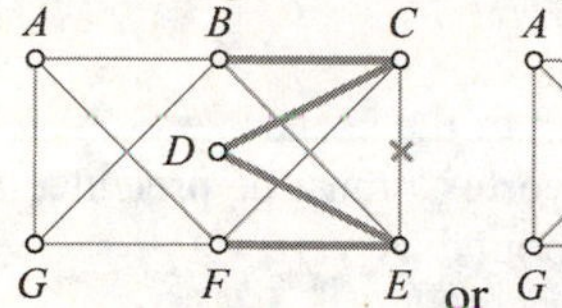

or

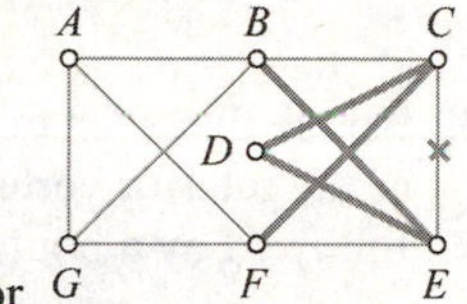

7. Strategy: Make an organized list. Suppose a Hamilton circuit starts by moving to the right (to vertex *B*). At *B* there are two choices as to where to go next. Each of these choices completely determines the rest of the circuit.

Starting by moving to the right will force the circuit to end at *F*. So the mirror image will represent the case of starting by moving left.

9. Any Hamilton path passing through vertex *A* must contain edge *AB*. Any path passing through vertex *E* must contain edge *BE*. And, any path passing through vertex *C* must contain edge *BC*. Consequently, any (Hamilton) path passing through vertices *A*, *C*, and *E* must contain at least three edges meeting at *B*. This is a problem (why?).

11. The existence of the bridge *BC* is really the root cause of the lack of Hamilton circuits and certain Hamilton paths in this problem. Why?

13. Don't work too hard on part (c). Remember that Hamilton circuits can be traveled forwards or backwards.

15. There are only two Hamilton paths that start at *A* and end at *C*.

B. Factorials and Complete Graphs

17. **(a)** Strategy: Look for a pattern.
$10! = 10\times(9\times8\times\ldots\times3\times2\times1)$

(c) Strategy: Look for a pattern.
$11! = 11\times10\times(9\times\ldots\times3\times2\times1)$

19. **(a)** Strategy: Look for a pattern.
$$\frac{9!+11!}{10!} = \frac{9!}{10!} + \frac{11!}{10!} = \frac{9!}{10\times9!} + \frac{11\times10!}{10!}$$

21. The factorial key on your calculator may be hidden under a menu (such as PRB for probability). If the number is large, the calculator will output the number in scientific notation. The two digits on the far right will represent the power of 10 in this notation.

23. Use the fact that there are 60 seconds each minute, 60 minutes each hour, 24 hours each day, and 365 days each year. Keeping track of units will help you with the calculation.

25. **(a)** Strategy: Work backwards.
In general, K_N has $N(N-1)/2$ edges.

(c) When one new vertex is added to K_{50} in order to form K_{51}, how many edges must also be added?

27. **(a)** Strategy: Work backwards.
The number of distinct Hamilton circuits in K_N is $(N-1)!$.

(b) Strategy: Work backwards.
K_N has $N(N-1)/2$ edges.

C. Brute Force and Nearest Neighbor Algorithms

29. **(a)** Strategy: Make an organized list.

31. Review the nearest-neighbor tour constructed in Example 6.7.

33. Use brute-force on (c). There are only six possible circuits that make *B* the first stop after *A*.

35. See Example 6.8.

D. Repetitive Nearest-Neighbor Algorithm

37. Strategy: Make a table or chart.
Essentially, you are applying the nearest-neighbor algorithm five times. This should feel tedious, but not complicated.

The nearest-neighbor tour found by starting at vertex A has weight 16.0. [This is one-fifth of the work to be done.]

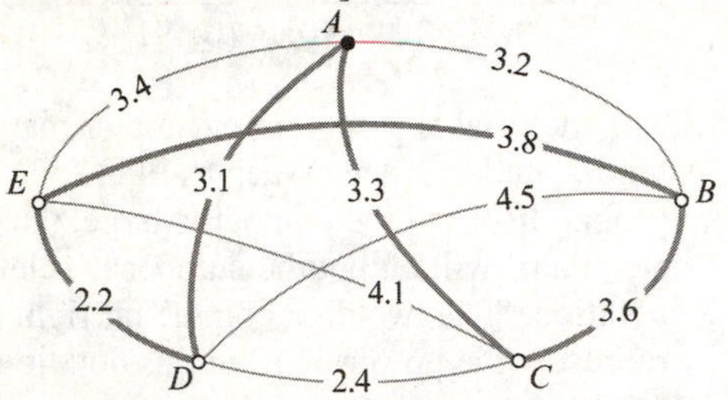

E. Cheapest-Link Algorithm

43. The first three steps of the cheapest-link algorithm are shown in the following figures.

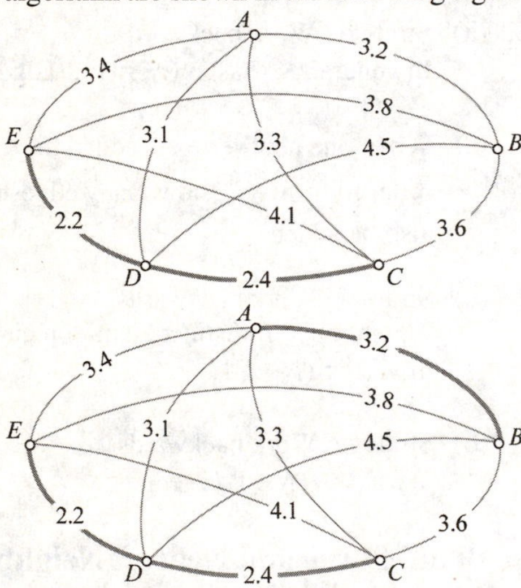

After the second edge CD is added, the next cheapest edge is 3.1. Since this edge makes three edges come together at vertex D, we skip adding this edge.

F. Miscellaneous

49. See Example 6.9.

51. The relative error of the nearest-neighbor tour is $\varepsilon = (W - Opt)/Opt$ where W is the weight of the nearest-neighbor tour and Opt is the weight of the optimal tour. Here W and ε are known.

53. Strategy: Draw a diagram.
The weighted graph will represent the six locations using vertices and the weights of the edges between each vertex will represent the distance (the number of blocks a taxi would need to drive) between the vertices.

55. A seating arrangement can be found from a Hamilton circuit.

JOGGING

61. **(a)** K_{501} has one more vertex and several more edges (one for each edge connecting to this "new" vertex) than K_{500}. The same can be said of the relationship between K_{502} and K_{501}.

63. Notice that each of the corner vertices as well as the interior vertex I must be preceded and followed by a boundary vertex.

65. Think of the vertices of the graph as being colored like a checker board (red and black). Why can't you start at a red vertex and end at a red vertex?

69. **(a)** Is crossing a bridge twice a problem? Why?

71. See Example 6.8.

Chapter 7

WALKING

A. Trees

1. **(a)** A tree is a network with no circuits. Properties 1-3 will help you decide if a graph is a tree or not.

(b) Remember that a network is a connected graph.

(d) Compare this graph with that found in (a). How different are these?

3. **(a)** See tree property 3.

(b) Use your answer to (a) and tree property 2.

5. **(a)** See tree property 3.

(c) See tree property 2.

(d) See tree property 1.

9. If all of the vertices are even, the graph has an Euler circuit.

B. Spanning Trees

11. (a) Many answers are possible. Each answer will consist of exactly 4 edges. Two vertices will have degree 1.

(b) Recall that if a network has *N* vertices and M edges, its redundancy is $R = M - (N - 1)$.

15. (a) Strategy: Make an organized list. There are three ways to eliminate the only circuit *D*, *I*, *E*, *D*.

(c) Strategy: Make an organized list. How many ways are there to eliminate the only circuit *D*, *I*, *J*, *E*, *K*, *L*, *D*?

17. (a) Each spanning tree excludes one of the edges *AB*, *BC*, *CA* and one of the edges *DI*, *IE*, *EJ*, *JD*. The multiplication rule will be useful here. You can review it in chapter 2.

(b) Again, use the multiplication rule.

C. Minimum Spanning Trees and Kruskal's Algorithm

19. (a) Add edge *EC* to the tree first. Then, add edge *AD*.

(b) The total weight is 165 + 185 + ….

23. Kansas City – Tulsa is the cheapest edge. Pierre – Minneapolis is next cheapest. Continue adding edges to the tree without forming a circuit.

25. How many edges will a spanning tree for a graph having 20 vertices have?

D. Steiner Points and Shortest Networks

27. First, determine the measures of each angle in the triangle if possible. If they are all less than 120°, then the network has a Steiner point. If not, the shortest network will be the MST.

29. (a) This triangle is isosceles.

(b) How does $m(\angle ABC)$ compare to $m(\angle ACB)$? What does that say about $m(\angle CAB)$?

31. Since $m(\angle CAB) = 128°$, there is no Steiner point. Find the length of *AB* and double it (why?) since the MST is the shortest network.

33. The Steiner point will create the shortest network. In fact, you can compute the length of that network from the table in this case.

35. (a) Could *E* be part of the shortest network connecting *B*, *C*, and *D*? Why or why not?

(b) Find $m(\angle CDB)$.

(c) $m(\angle CEB) > 120°$

E. Miscellaneous

37. (a) The network could have no circuits,

or a circuit of length 3,

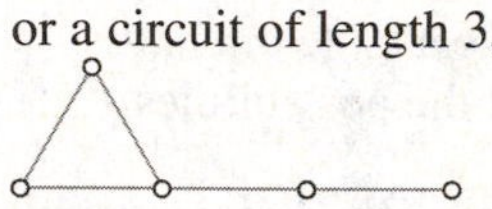

or …

(b) Strategy: Look for a pattern. There is a pattern in part (a).

39. (a) See property 4.

(b) How many edges will need to be "discarded" in order to form a spanning tree?

(c) Are any edges in the circuit bridges? How about the edges not in the circuit?

43. (a) Imagine that the figure is the top half of an equilateral triangle.

(b) By the Pythagorean theorem, $h^2 = s^2 + l^2$. Then, use the result of (a).

45. $\triangle CSB$ is made up of two side by side 30°-60°-90° triangles (why?). Use the result of exercise 43(a) to determine the length of *CS* (which is the same as the length of *BS*).

47. Use the results of exercise 43.

49. In a 30°-60°-90° triangle, the side opposite the 30° angle is one half the length of the hypotenuse and the side opposite the 60° angle is $\sqrt{3}/2$ times the length of the hypotenuse (see exercise 43). Also, $m(\angle BAC) = 60°$.

JOGGING

51. **(a)** Strategy: Make an organized list. See Examples 7.4 and 7.5.

(b) Since the circuits *A*,*B*,*C*, *D*,*E*,*F*,*A* and *E*,*I*,*H*,*G*,*F*,*E* share a common edge *EF* there are two ways to exclude edges to form a spanning tree. If one of the excluded edges is the common edge *EF*, then any one of the other 9 edges could be excluded to form a spanning tree. If, on the other hand, one of the excluded edges is not the common edge *EF*, then one excluded edge has to be from the "top circuit" and the other excluded edge must be from the "bottom circuit."

55. If $M \geq N$, what can be said about the redundancy of the network?

57. Strategy: Make a table or chart. Exhaust all of the possibilities.

61. **(a)** $\angle BFA$ and $\angle EFG$, for example, are supplementary angles.

(b) Show that the angles in triangle *ABC* are all measure less than 120°.

(c) Show that a Steiner point *X* could not fall in triangle *ABF*, triangle *BCG*, or triangle *ACE*. That is, show that the angles that *X* would form in those circumstances are not all 120°.

63. **(a)** The length of the network is $4x + (300 - x)$.

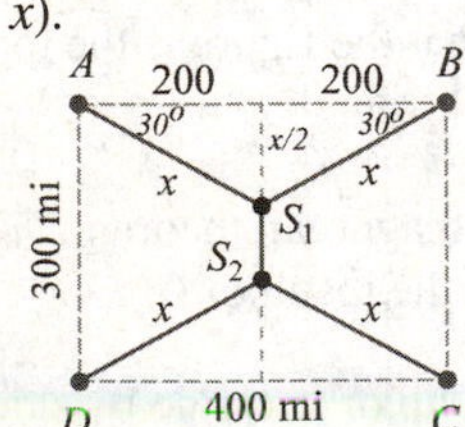

Chapter 8

WALKING

A. Directed Graphs

1. **(a)** Remember that the arcs are directed so that *AB* is not interpreted in the same way as *BA*.

(b) How many arcs are aimed at each vertex? How does the sum of these indegrees compare to the number of arcs in the arc-set?

3. **(a)** See Exercise 1(a).

(b) This can be determined without counting one by one.

5. **(a)** If *XY* is an arc in the digraph, we say that vertex *X* is *incident to* vertex *Y*.

(b) If *XY* is an arc in the digraph, we say that vertex *Y* is *incident from* vertex *X*.

(e) The arc *YZ* is said to be *adjacent to* the arc *XY* because the starting point of *YZ* is the ending point of *XY*.

7. Many answers are possible. Start by drawing the vertices. Then, draw each arc one at a time.

9. The outdegree of *X* is the number of arcs that have *X* as their starting point (outgoing arcs); the indegree of *X* is the number of arcs that have *X* as their ending point (incoming arcs).

11. **(a)** Remember that a path must follow in the direction of the arcs.

(b) See chapter 6.

(c) A cycle must follow the direction of the arcs. It is a path that starts and ends at the same vertex.

(d) If *F* were part of a cycle, what vertex would be passed to after *F*?

(f) Strategy: Draw a diagram. Sometimes having a visual representation is helpful.

15. **(a)** Is there anyone that everyone respects?

(b) Is there anyone that no one respects?

B. Project Digraphs

17. Try putting tasks that need very little completed before they can start on the left and those that need a lot of other tasks completed before they can start on the right.

19. Strategy: Make a table or chart. Putting the information in a chart such as that given in exercise 17 would be helpful.

C. Schedules and Priority Lists

23. **(a)** There are $31 \times 3 = 93$ processor hours available. How many of these will be used?

(b) How long would it take if there were no idle time?

25. The best case scenario is that the work is divided evenly among all six processors.

27. At time $T = 26$, task *IW* is being processed (See Figure 8-14). However, no other task is ready to be processed (and so the other processor is idle).

29. The first processor would start on task *C* (since task *D* is not ready). The second processor would begin on task *A*. Once task *A* is completed, the next ready task is *B* (task *E* is not quite ready). After completing task *C*, the first processor would start task *E*.

33. The general cleaning requires quite a few other tasks to be completed before it is started.

35. This schedule will have quite a bit of idle time for the second worker while the first worker is painting. Let's hope you are the second (and not the first) worker in this scenario.

D. The Decreasing-Time Algorithm

37. **(a)** The decreasing-time list is formed by listing the tasks from longest to shortest. Clearly, *D*(12) is the longest task and will occur first in the list.

(b) The first processor starts with task *C* and the second processor starts with task *A*.

41. **(a)** If there were three processors, scheduling would be pretty easy and $Fin = 4$ would happen. But with only two processors, the finishing time doubles.

(b) Strategy: Make an educated guess and check. Essentially a little shuffling of the result in (a) should do the trick. $Opt = 6$.

E. Critical Paths and the Critical-Path Algorithm

45. **(a)** The critical time for *F* is $1 + 0 = 1$. The critical time for *G* is $2 + 0 = 2$. The critical time of *D* is $12 + 1 = 13$. The critical time for *E* is $6 + 2 = 8$.

(b) You may be able to eyeball this. It is either a path on the top (*START, A, D, F, END* or *START, B, D, F, END*) or the path on the bottom of the digraph.

(d) What is the best you can do with two workers (that must complete 43 hours worth of work)?

47. The critical path list is: *B*[46], *A*[44], *E*[42], *D*[39], *F*[38], *I*[35], *C*[34], *G*[24], *K*[20], *H*[10], *J*[5]. Here is a start to the schedule:

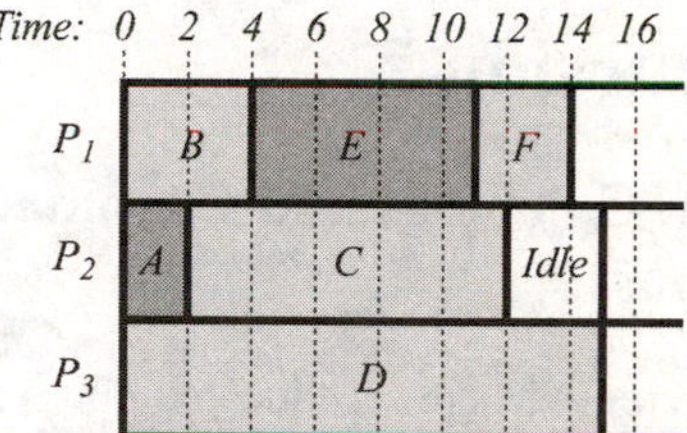

49. The project digraph, with critical times is:

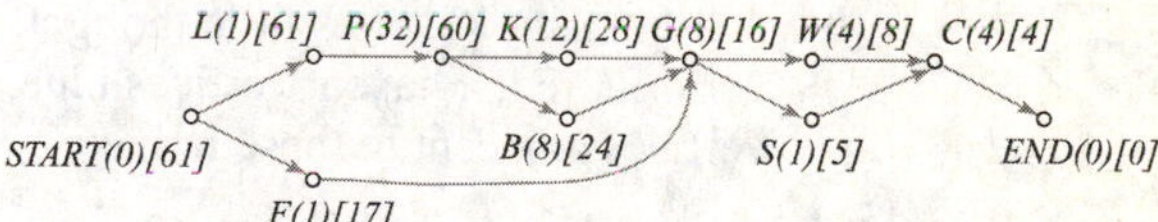

F. Scheduling with Independent Tasks

51. **(a)** Remember that the critical-path algorithm is the same as the decreasing-time algorithm when the tasks are independent.

(b) The answer in (a) looks pretty good.

(c) Not much error here.

53. See exercise 51.

55. **(a)** Since all tasks are independent, the critical-time priority list is identical to a decreasing-time priority list.

(b) $Opt = 12$. Finding the schedule is like putting a small jigsaw puzzle together.

57. **(a)** This schedule fits together very nicely, thank you. However, it should be noted that one processor focuses on one task.

G. Miscellaneous

59. See the hint to exercise 9.

JOGGING

67. **(a)** This is simply a matter of evaluating the formula for these values of N. It may help to convert your answers to percentages.

(b) $\frac{N-1}{3N} \leq \frac{N}{3N}$

73. **(a)** It may help to think about evenly dividing the work to be done.

(c) $N \times Fin$ is relevant as far as how much you need to *pay* the workers.

mini-excursion 2

A. Graph Colorings and Chromatic Numbers

1. **(a)** Strategy: Make an educated guess and check. To that end, you will find it useful to use an applet available in MyMathLab.

(b) A 2-coloring requires little thought. Color vertex A red, adjacent vertices blue, vertices adjacent to these red, etc.

(c) The only graphs having 1-colorings are not very interesting (no edges).

7. **(a)** In a compete graph K_n, pick a vertex. Any vertex. To how many vertices is that vertex adjacent?

(b) If we remove one edge from K_n, then there are two vertices that are not adjacent. They can be colored with the same color (say blue).

9. **(a)** Try coloring by traveling clockwise around the circuit. Be as efficient as possible along your route.

(b) Use the same approach as in (a). What happens at the end of your trip? [In case you don't see it, try the case of $n = 5$.]

11. Color one vertex v on the tree blue. Pick any other vertex on the tree. Can you figure out what color it will be? Does this tell you how to efficiently color the tree?

B. Map Coloring

13. For a simple example, think of a bunch of states shaped like Colorado and Wyoming.

15. **(a)** In the dual graph, each vertex represents a country in South America and two vertices are connected by an edge if the corresponding countries have a common border. Get your globe out.

(b) In the list of vertices, Brazil will be first. So it gets any color you wish. Bolivia, Argentina, and Peru could all be next on the list.

C. Miscellaneous

17. **(a)** No vertices in A are connected to each other.

(b) Strategy: Draw a diagram. Say the vertices are colored blue and red. Redraw the graph so that the blue vertices are all in a row on the top and the red ones are all in a row on the bottom. What do you notice?

(c) See exercise 9 (if a circuit were to have an odd number of vertices).

19. **(a)** See Euler's Sum of Degrees theorem in chapter 5.

(b) Use exercise 8 to show that $\chi(G)$ cannot be 1 and exercise 17 to see that $\chi(G)$ cannot be 2.

21. Suppose that the graph G is colored with $\chi(G)$ colors: Color 1, Color 2, ..., Color K. Now make a list $v_1, v_2, \ldots, v_n$ of the vertices of the graph as follows: All the vertices of Color 1 (in any order) are listed first (call these vertices Group 1), the vertices of Color 2 are listed next (call these vertices Group 2), and so on, with the vertices of Color K listed last (Group K).

Chapter 9

WALKING

A. Fibonacci Numbers

1. (a) The subscript indicates its place in the sequence. Essentially you are being asked to compute the 10th term in the Fibonacci sequence 1,1,2,3,...

(b) Add 2 *after* finding the value of F_{10}.

(c) $10 + 2 = 12$

(d) Divide by 2 *after* finding the value of F_{10}.

(e) $10/2 = 5$

3. (a) This is a sum of 5 Fibonacci numbers.

(b) $1 + 2 + 3 + 4 + 5 = 15$

(c) Order of operations alert: Subscripts take precedence over multiplication.

(d) $3 \times 4 = 12$

(e) Spend some time trying to understand the problem. Note that $F_4 = 3$.

5. (a) F_N represents the *N*th Fibonacci number. What is being done to that number?

(b) F_{N+1} represents the Fibonacci number in position $(N + 1)$.

(c) F_{3N} represents the Fibonacci number in the 3*N*th position.

(d) The 3*N*+1 is computed first in evaluating this expression.

7. (a) Fibonacci numbers can be found recursively by $F_N = F_{N-1} + F_{N-2}$.

(b) Strategy: Work backward.
Add two Fibonacci numbers to "move forward" in the Fibonacci sequence.

9. You are looking for a way to express the fact that each term of the Fibonacci sequence is equal to the sum of the two preceding terms.

11. (a) Strategy: Make an educated guess and check. 47 is the sum of just two Fibonaccis.

(b) Strategy: Make an educated guess and check. You'll need three Fibonaccis here. Try using numbers as large as possible.

(c) Try starting with 144 as one summand. Use the largest Fibonacci number available at each stage of constructing your sum.

(d) Again, 144 is the largest Fibonacci number less than 210. Try using that in your sum. Think efficiency.

13. (a) Strategy: Look for a pattern.
The right hand side appears to be the Fibonacci number that follows that last term on the left hand side.

(b) Strategy: Look for a pattern.
The subscripts in the fourth equation are 1, 3, 5, 7, 9, and 10. Do you see a pattern? Also note that when the left hand side of the fourth equation ended in F_9, the right hand side was F_{10}.

(c) Strategy: Look for a pattern.
The subscript on the right hand side compares closely to the subscript on the last term of the sum.

15. (a) Try the Fibonacci numbers $F_4 = 3$, $F_5 = 5$, $F_6 = 8$, and $F_7 = 13$.

(b) Denoting the first of these four Fibonacci numbers by F_N, the second of these four Fibonacci numbers is F_{N+1}, the third is F_{N+2}, and the fourth is F_{N+3}.

B. The Golden Ratio

19. (a) The order of operations is critical. First add 1 and $\sqrt{5}$. Next, divide by 2. Then multiply by 21. Finally, add 13.

(b) Order of operations is again critical. First add 1 and $\sqrt{5}$. Next, divide by 2. Finally, raise the result to the 8th power.

(c) First add 1 and $\sqrt{5}$. Next, divide by 2. Take the result to the power of 8. Put the result in memory. Next, subtract $\sqrt{5}$ from 1. Then, divide that by 2. Take the resulting number to the power of 8 (as before). Subtract this number from what you have in memory. Finally, divide what you have computed so far by $\sqrt{5}$.

23. (a) Use the following facts: $\phi^6 = \phi^5 \cdot \phi$ and $\phi^2 = \phi + 1$.

(b) Replace ϕ with $\frac{1+\sqrt{5}}{2}$ and simplify (collect like terms).

25. The hint also says that $F_N \approx \phi \times F_{N-1}$. Be sure to convert the result of any calculations to scientific notation.

27. (a) To find the value of A_7, double the 6th term and add the 5th term. That is,

$$A_7 = 2A_{7-1} + A_{7-2}$$
$$= 2A_6 + A_5$$

(b) That is, estimate the value of $\frac{A_7}{A_6}$.

(c) To find $\frac{A_{11}}{A_{10}}$, you will need to find the first 11 terms of the sequence using the given recursive formula.

(d) Just like ratios of successive Fibonacci numbers settle down to a magical number (in that case the golden ratio), so do ratios of successive terms in the Fibonacci sequence of order 2.

C. Fibonacci Numbers and Quadratic Equations

29. (a) Rewrite this (quadratic) equation as $x^2 - 2x - 1 = 0$. Here a = 1, b = -2, and c = -1. So, by the quadratic formula,

$$x = \frac{-(-2) \pm \sqrt{(-2)^2 - 4(1)(-1)}}{2(1)}.$$

31. (b) Rewrite the equation:

$55x^2 - 34x - 21 = 0$

Then a = 55, b = -34 so that the sum of the two solutions is -b/a = 34/55.

33. (a) What does it mean for x = 1 to be a "solution" to an equation?

(b) Rewrite the equation as $F_N x^2 - F_{N-1}x - F_{N-2} = 0$.

D. Similarity

35. (a) Strategy: Draw a diagram. Since R and R' are similar, each side length of R' is 3 times longer than the corresponding side in R.

(b) Strategy: Solve a simpler problem first. Suppose that R and R' were squares. How many times larger would the area of R' be in that case?

37. (a) The ratio of the perimeters must be the same as the ratio of corresponding side lengths.

(b) For two similar triangles T and T', if the side lengths of T' are x times larger than the corresponding sides in T, then how does the area compare? [Note: The same fact holds true for rectangles and squares.]

39. Strategy: Draw a diagram. There are two possible cases that need to be considered. First, we might solve $\frac{3}{x} = \frac{5}{8-x}$. But, it could also be the case that the side of length 3 does not correspond to the side of length 5, but rather the side of length 8 – x. In this case, we solve $\frac{3}{x} = \frac{8-x}{5}$. So, more than one solution is possible.

E. Gnomons

41. So 3 is to 9 as 9 is to c+3.

43. 8 is to 12 as 1+8+3 is to 2+12+x.

45.

	A	B
20	10	x

10 + x

47. (a) Angle CAD and angle BAC are supplementary. To determine the measure of angle BDC, use the fact that triangle BDC must be similar to triangle BCA (in order for triangle ACD to be a gnomon). Finally, use the fact that the sum of the measures of the angles in any triangle is 180°.

(b) Use the fact that triangle *DBC* is isosceles.

49. Strategy: Draw a diagram.
It helps to draw the original figure *F*, the gnomon *G*, and the joining of the two. The legs of the two similar triangles must be proportional. So, 3 is to 4 as 9 is to *x*.

JOGGING

51. Strategy: Look for a pattern.
The terms in the sequence given appear to have a common factor.

53. **(a)** $T_1 = 7F_2 + 4F_1$ and $T_2 = 7F_3 + 4F_2$

(b) Notice that

$$T_{N-1} + T_{N-2} = (7F_N + 4F_{N-1}) + (7F_{N-1} + 4F_{N-2})$$

Write this value in terms of F_{N+1} and F_N.

(c) You are being asked to describe T_1, T_2, and the value of T_N in terms of its predecessors.

55. Strategy: Solve a simpler problem first.
As a similar example, the irrational numbers $\frac{5}{3} = 1.\overline{6666}$ and $-\frac{2}{3} = -0.\overline{6666}$ sum to an integer 1. So, they have the same decimal expansion.

57. Use the fact that $F_N\phi + F_{N-1} = \phi^N$.

59. If the rectangle is a gnomon to itself, then the rectangle in the figure below must be similar to the original *l* by *s* rectangle.

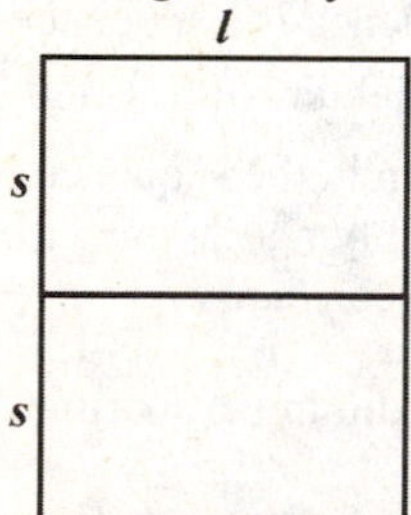

Use this to determine the value of *l/s*.

61. Since the area of the white triangle is 6, the area of the shaded figure must be 48, which makes the area of the new larger similar triangle 6+48=54. Since the ratio of the areas of similar triangles is the square of the ratio of the sides, we have, for example,

$$\frac{3+x}{3} = \sqrt{\frac{54}{6}} = 3.$$

63. The ratio of the area of similar rectangles is the square of the ratio of the sides.

65. **(a)** Show that triangle *ABD* is isosceles with $AD = AB = 1$. Therefore $AC = x - 1$. Using these facts and the similarity of triangle *ABC* and triangle *BCD* we have

$$\frac{x}{1} = \frac{1}{x-1}.$$

(b) The fact that every triangle in sight is isosceles in this problem should help.

67. **(a)** Split the regular decagon into ten equally sized triangles all having a vertex at the center. These triangles are each isosceles 36-72-72 triangles. Exercise 65 tells you all you need to know about such triangles to compute the perimeter.

(b) All ten triangles in this part of the problem have side lengths *r* times larger than in the previous part.

Chapter 10

WALKING

A. Percentages

1. **(a)** "per cent" means per 100 so 2.25% is equivalent to 2.25 per 100.

(b) Drop the % sign and move the decimal point two places to the left

3. **(a)** 2.25 out of every 100 is the same as 225 out of every 10,000.

5. 60 is to 75 as what is to 100?

7. Strategy: Work backwards.
If 14% of the pieces are missing, then 86% of the pieces are present.

9. Strategy: Use a variable.
$\$6.95 + tax = \$7.61$; But, $tax = tax\ rate \times \$6.95$. If we call the unknown tax rate *x*, then $\$6.95 + x \cdot \$6.95 = \$7.61$.

11. 16% is to \$2 trillion as 100% is to what?

13. If T was the starting tuition, then the tuition at the end of one year is 110% of T. That is, $(1.10)T$. The tuition at the end of two years is 115% of what it was after one year. That is, after two years, the tuition was $(1.15)(1.10)T = 1.265T$. That's a 26.5% increase over the course of two years.

15. Suppose that the cost of the shoes is originally \$$C$. After the 120% mark up, the price is \$$2.2C$. When the shoes go on sale, they are priced at 80% of that price or \$$(0.8)(2.2)C$. Apply the other discounts in the same way to figure out what percent of the original price the final price is. This will spell out the final percentage of profit.

17. Suppose that when the week began the DJIA had a value of A. At the end of Monday, it was valued at $1.025A$. At the end of Tuesday, it was valued at $(1.121)(1.025)A = 1.149025A$ (an increase of 14.9025%).

B. Simple Interest

21. (a) Apply the Simple Interest Formula to find the future value F for $P = \$875$, $r = 0.0428$, and $t = 4$.

23. Determine how much the bond is worth after five years. Then, calculate how much interest in earned. Since 85% of that is what you keep, calculate 85% of the interest to find how much money you net in the end.

25. Strategy: Work backwards.
Apply the Simple Interest Formula to find the present value P for $F = \$6000$, $r = 0.0575$, and $t = 4$.

27. Strategy: Work backwards.
Apply the Simple Interest Formula to find the APR as a decimal (r) given $P = \$5400$, $F = \$8316$, and $t = 8$.

29. You guessed it – use the Simple Interest Formula. In this case, $F = 2P$.

C. Compound Interest

31. (a) Apply the Annual Compounding Formula to find the future value F when $P = \$3250$, $r = 0.09$ and $t = 4$.

(b) Since interest is credited to the account at the end of each year, no growth takes place during the last seven months. Depositing the money for 5 years and 4 months would produce the same result.

33. Determine how much is in the account on Jan. 1, 2011. Then, use that amount to determine the balance on Jan. 1, 2015.

37. (a) The periodic interest rate p is the annual interest rate r divided by the number of periods n in a year. And, in case you forgot, there are 12 months (the period in this exercise) in a year. The \$5000 grows for 5 years (T = ? months) at the periodic rate p. See version 2 of the General Compounding Formula.

(b) Each \$1 invested grows to

$$\$\left(1+\frac{0.12}{12}\right)^{12} \approx \$1.126825 \text{ after one year.}$$

What percent return on \$1 is this?

39. (a) First, figure out how many periods (hours) there are in one year. Then, apply the General Compounding Formula using that value for n.

45. The APY is the percentage of profit that an investment generates in a one-year period. The profit in this case is clearly \$810 - \$750.

47. (a) Strategy: Make an educated guess and check.

49. Strategy: Work backwards.
Use the General Compounding Formula and solve for the initial principal P.

D. Geometric Sequences and The Geometric Sum Formula

51. (a) A geometric sequence is a sequence that starts with an initial term (G_0) and from then on every term in the sequence is obtained by multiplying the preceding term by the same constant c.

(b) Repeat what you did in (a) six times.

(c) Strategy: Look for a pattern.
Repeat what you did in (a) N times.

53. (a) Strategy: Work backwards.
Use the fact that $G_1 = c \cdot G_0$ to determine the value of the common ratio c.

55. (a) Let G_N represent the number of crimes committed in the year 2000 + *N*. Multiplying the number of crimes by 1.50 each year will account for a yearly *increase* by 50%.

57. (c) Since this is the sum of terms in a geometric sequence, you should apply The Geometric Sum Formula.

(d) Strategy: Solve a simpler problem first. Again, you can use The Geometric Sum Formula, but this time the value of *P* (the first term) is different (it is 3×2^{50}). There are $N = 51$ terms in the sum.

61. Apply The Geometric Sum Formula with $P = \frac{10}{1.05}$ and $c = \frac{1}{1.05}$.

E. Deferred Annuities

63. Apply the Fixed Deferred Annuity Formula with $T = 40$ payments (deposits) of $\$P = \$2000$ having a periodic (yearly) interest rate of $p = 0.075$. The future of the value of the last payment, *L*, is the amount that a *single* deposit grows to in one year.

65. Strategy: Solve a simpler problem first. Compare this exercise with 63. Note that the only difference is that the payments are made at the end of each period. So, the value of *L*, the future value of the last payment, will differ.

67. The periods are months and the 11 payments are made at the beginning of each. The monthly interest rate is 1/12 of the yearly rate.

69. Strategy: Work backwards. First, figure out the future value of Celine's account at the end of 3 years (36 months). Then, since this is just a down payment, calculate how much house she can afford.

71. Strategy: Work backwards. Solve for $\$P$ in the Fixed Deferred Annuity Formula where $F = \$150{,}000$, $T = 12\times52 = 624$ and $p = 0.0468/52 = 0.0009$.

F. Installment Loans

73. (a) Apply the Amortization Formula with $T = 25$ payments of $\$F = \$3000$ having a periodic interest rate of $p = 0.06$. Note: $q = \frac{1}{1.06}$.

75. The periods are weeks and the periodic interest rate is $p = 0.052/52 = 0.001$.

77. (a) Solve for *F* in the Amortization Formula with $T = 30\times12 = 360$ payments and a monthly interest rate of $p = 0.0575/12$.

(b) The amount of interest paid is the number of payments times the size of each payment (i.e. the total paid over all of these years) less the amount of the original loan.

79. Find the present value of this installment loan and then add $25,000.

JOGGING

81. Strategy: Solve a simpler problem first. The answer is NOT 75%. Try it with goods that are worth $1 for starters.

83. (a) How would *y*% be represented without the % sign? How is that expression related to a *y*% discount? Remember, you pay 100% less the percentage discount.

85. Your answer will be reasonably close to $600 million.

Chapter 11

WALKING

A. Reflections

1. (a) Upon reflection, the image of *P* with axis l_1 is found by drawing a line through *P* perpendicular to l_1 and finding the point on this line on the opposite side of l_1 which is the same distance from l_1 as the point *P*.

(c) Since l_3 has slope of 1, the line perpendicular to it has a slope of -1 (the negative reciprocal. That is, the line perpendicular to l_3 will be parallel to l_4.

(d) Since l_4 has slope of -1, the line perpendicular to it has a slope of 1.

3.

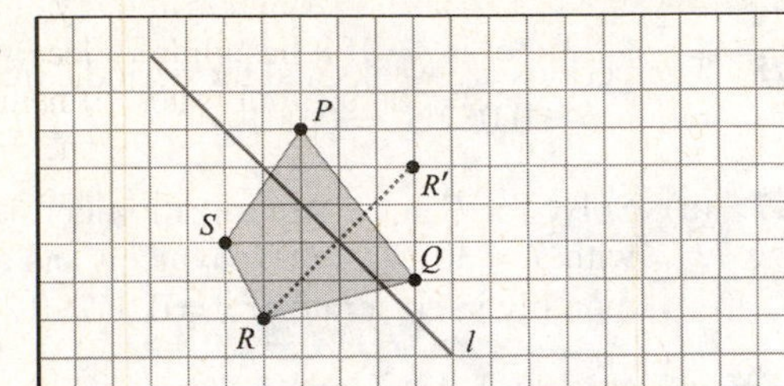

The image of *R* can be found by drawing a line perpendicular to *l* and finding the point *R'* on the opposite side of this line which is the same distance from *l* as the point *P*.

5. (d) A fixed point of a rigid motion is a point that is not moved by that motion.

7. (a) The line segment *PP'* has a slope of -1 (since moving from *P* to *P'* requires one to move down 7 units and right 7 units). So, the axis of reflection will have a slope of 1 and will pass through the midpoint of segment *PP'*.

9. A fixed point of a rigid motion is a point that is not moved by that motion. For a reflection, these will always occur on the axis of reflection.

B. Rotations

11. (a) Strategy: Create a physical or graphical model. Think of *A* as the center of a clock in which *B* is the "9." Where is the "12" of this clock located?

(b) Think of *B* as the center of a clock in which *A* is the "3."

(c) Think of *B* as the center of a clock in which *D* is the "1." Where is the "3" located (60° is 1/6, or 2/12, around the clock)?

(d) 120° is 1/3, or 4/12, of the distance around a circle.

(e) Think of *A* as the center of a clock in which *I* is the "12." Rotating 3690° has the same effect as rotating several times all of the way around the clock (plus a remainder). In the remainder lies the answer.

13. (a) Imagine starting at "12" on a clock and rotating clockwise 710° . How far would you need to rotate the "12" in the other direction in order to end up at the same location? [There are 360° in a full circle.]

(b) This is almost two full trips around a circle.

(c) $360^\circ \overline{)7100^\circ}$ with quotient 19 R 260°

(d) $360^\circ \overline{)71{,}000^\circ}$ with quotient 197 R 80°

15. Since *BB'* and *CC'* are parallel, the rotocenter will be located at the intersection of the perpendicular bisector of these segments and the line that passes through *BC* (see Figure 11-9(c)).

17. The rotocenter *O* is located at the intersection of the perpendicular bisectors to *PP'* and *SS'*. Since *PP'* has a slope of -1, then its perpendicular bisector will have a slope of 1. Since *SS'* has a slope of -2, its perpendicular bisector will have a slope of ½. Naturally, you should also remember that a perpendicular bisector of a segment passes through the midpoint of that segment.

19. The grid consists of equilateral triangles. Each angle in such a triangle measures 60° . Try rotating the point *A* first, since it is easiest.

C. Translations

21. (a) Vector v_1 translates a point 4 units to the right.

(b) Vector v_2 also translates a point 4 units to the right.

23. Each point moves in the same direction and the same distance under the translation. The location of *E'* unlocks the secret of how each point will move.

25. Don't let the grid distract you. The answer is the same with or without the grid.

D. Glide Reflections

27. Under a glide reflection, you may glide first and then reflect or reflect first and then glide. It's all up to you.

29. The axis of reflection for this glide reflection passes through the midpoints of *BB'* and *DD'*. Once you determine it, reflect first (to find a figure *A*B*C*D*E** and then glide from there to the final destination called *A'B'C'D'E'*.

33. Locating the midpoints of *CC'* and *DD'* isn't really any different than it would be on a square grid. Since *DD'* are on the same line, the midpoint is easy to spot. To locate the midpoint of *CC'*, imagine that you were traveling from *C* to *C'* along the gridlines. One efficient path might be moving to the left two units and then SSW six units. So, the midpoint would be located after traveling one unit left and three units SSW. After the axis of reflection is found, remember that a reflection is perpendicular to this axis (the different grid should not confuse you).

E. Symmetries of Finite Shapes

35. Strategy: Create a physical or graphical model. If you are really stuck on these, try cutting out these shapes with a pair of scissors and fold (reflect) and rotate to discover the symmetries of each.

39. Start by determining if there are any reflectional symmetries. If so, it is of type *D*. If not, then it is of type *Z*. The number of rotations the object has determines the value of the subscript.

41. Count reflectional symmetries. If there aren't any, then count rotational symmetries.

45. Strategy: Make an educated guess and check. Experiment a bit. Keep your letters "blockish" so that they have the chance at having symmetry (e.g. A has vertical symmetry but A does not).

47. Answers abound. See how creative you can be.

F. Symmetries of Border Patterns

49. Here is an algorithm you can use: To determine the first symbol (*m* or 1), determine whether the pattern has vertical (*m*idline) symmetry. If so, the first symbol is an *m*, if not, it is a 1.

To determine the second symbol (*m*, *g*, 2, or 1), determine where the pattern has horizontal (*m*idline), *g*lide reflection, or half-turn (1/2) symmetry. If it has horizontal symmetry, the second symbol is *m*. If it has glide reflectional symmetry, the second symbol is a *g*. If neither of those apply and it has half-turn symmetry, the second symbol is a 2. If none of the above apply, the second symbol will be a 1.

53. Strategy: Solve a simpler problem first. Try drawing a object having this symmetry type and then form a border pattern by repeating it in the horizontal direction. That is, develop an example or two of your own.

G. Miscellaneous

55. (a) Only two of the rigid motions are proper. Only two of the rigid motions have any fixed points. This should narrow your search.

57. (a) The reflection of *P* about l_1 is the point *E*. What is the reflection of *E* about line l_2 ?

(b) Do what you did in (a) only in the opposite order.

59. (a) Will the left-right and clockwise-counterclockwise orientations on the final figure be the reverse of the original figure? Perhaps an experiment will help.

(c) This is an improper rigid motion combined with a proper rigid motion (see part (a)).

61. How many fixed points does the product have?

JOGGING

63. (a) Strategy: Draw a diagram. The diagram below shows a reflection from *P* to *P'* and then from *P'* to *P''*.

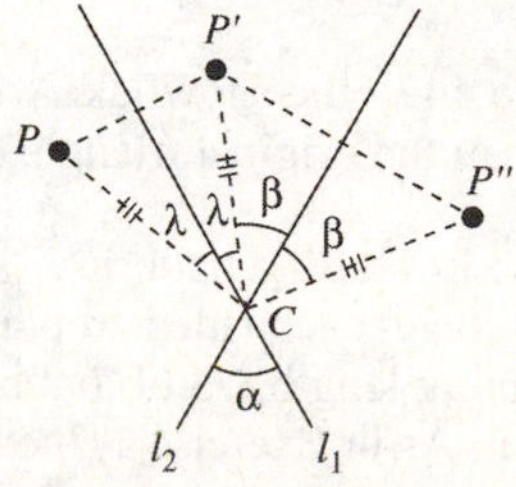

(b) Relabel the above diagram so that *P''* is *P* and *P* is *P''*.

67. **(a)** First find the image of the triangle under the translation. Then, take that image and rotate it.

(b) Is the product of these motions a proper motion? Does the product of these motions have a fixed point?

69. Remember that the only directions for translation symmetries in a border pattern are horizontal.

71. First, narrow the search down to an improper rigid motion since $\mathcal{M}$ is certainly that. What would have to be true about segments PP', RR', and QQ' if this were a reflection?

Chapter 12

WALKING

A. The Koch Snowflake and Variations

1. **(a)** At each stage, each side is replaced by 4 sides having length 1/3 of that being replaced. The length of each side times the number of sides gives the length of the boundary at each stage of construction.

(b) Strategy: Look for a pattern.
Look for a pattern in the first two columns of the table in part (a). For *P*, realize that it is the product of *M* and *l* at each stage of construction.

3. **(a)** Strategy: Look for a pattern.
At step 1, three triangles are added each having 1/9 the area of the original triangle. That is, the *increase* in area at step 1 is 27 in^2. At step 2, $3 \times 4 = 12$ triangles are added each having $\frac{1}{9^2}$ the area of the original triangle.

(b) See page 442 - the snowflake will be 60% larger than the original triangle.

5. **(a)** Strategy: Look for a pattern.
At each stage, each side is replaced by 5 sides having length 1/3 of that being replaced. As in Exercise 1, the length *l* of each side times the number of sides *M* gives the length *P* of the boundary at each step of construction.

(b) Strategy: Look for a pattern.
Once you have done step 30 in part (a), this isn't much different.

7. At step 1, four squares are added each having 1/9 the area of the original square. That is, the *increase* in area at step 1 is 36 in^2. At step 2, $4 \times 5 = 20$ squares are added each having $\frac{1}{9^2}$ the area of the original square. This results in an *increase* in area (from step 1) of $\frac{20 \times 81}{9^2} \text{ in}^2$.

9. Note the similarity in constructing the perimeter of the Koch antisnowflake with the Koch snowflake.

11. **(a)** At step 1, three triangles are subtracted each having 1/9 the area of the original triangle. That is, the *decrease* in area at step 1 is $T = \frac{3 \times 81}{9} \text{ in}^2$. At step 2, $3 \times 4 = 12$ triangles are subtracted each having $\frac{1}{9^2}$ the area of the original triangle (i.e. $S = 1$ sq. in. in step 2).

15. Note that for every square that is added, one like it is taken away.

B. The Sierpinski Gasket and Variations

17. Strategy: Look for a pattern.
In the step 1, $R = 1$ triangle is subtracted. At each step thereafter, the number of triangles subtracted is tripled. Starting with step 2, the area *S* of each triangle subtracted at a given step is 1/4 of the area of each triangle subtracted during the previous step (in step 1, the area of the triangle that is subtracted is 1/4).

19. At each step, the value of the number of dark triangles *U* is tripled and the value of *V* (the perimeter of each dark triangle) is halved (since each side length is halved).

21. The area of the Sierpinski gasket is smaller than the area of the gasket formed during any step of construction. Also note that if the area of the original square is 1, then the area at the *N*th stage of construction is $(3/4)^N$.

23. (b) One of the nine equally sized subsquares has been removed. That means that 8 of the 9 subsquares (and that fraction of the area) remains.

(c) Do what you did in (a) again.

25. (a) Each removed square will have sides of length 1/3 that of the square from which it was removed. At step 1, the boundary consists of the previous boundary plus the boundary of a new hole with sides of length 1/3. That is the length of the boundary has increased by $4\times(1/3)$. At step 2, there are 8 new holes, each with sides of length $\frac{1}{3}\times\frac{1}{3}=\frac{1}{9}$. This will *increase* the length of the boundary by $8\times\frac{4}{9}$.

(b) Determine the length of the boundary in step 1 and add the perimeter of 8 new square holes, each having sides of length $(1/3)^2=1/9$.

(c) Determine the length of the boundary at step 2 and add the perimeter of 8^2 new square holes, each having sides of length $(1/3)^3=1/27$.

27. The area *T* of the gasket subtracted at a given step of the construction will be the product of the number of triangles *R* subtracted at that step (6 times more than the previous step) and the area of each such triangle (1/9 of the area *S* of each triangle in the previous step).

29. Each dark triangle at a given step is replaced by 6 dark triangles. These smaller dark triangles each have perimeter of 1/3 the previous dark triangle's perimeter.

C. The Chaos game and Variations

33. Start at the first "winning" vertex P_1 = (32, 0). Then, move halfway to the next winning vertex *A* (that is, find the midpoint of *A* and P_1). This will put you at P_2 = (16, 0). Next, move halfway to C, the next winning vertex (i.e. find midpoint of *C* and P_2).

37. Find the new coordinates by picking new *x*-value to be 2/3 of the way from *x*-coordinate of first point to *x*-coordinate of second point and picking new *y*-value to be 2/3 of way from *y*-coordinate of first point to *y*-coordinate of second point.

So, starting at P_1 = (0, 27), move 2/3 of the way to the next winning vertex *B* = (27,0). x = 18 is 2/3 from P_1 to *B*. $y = 9$ is 2/3 from P_1 to *B* (the *y*-coordinate is moving down).

39. (a) Strategy: Work backwards.
P_1 = (0,27) corresponds to a first roll of 4. Then, since P_2 = (18,9) is 2/3 of the way from P_1 to *B*, the second roll was 2.

D. Operations with Complex Numbers

41. (a) Remember that $i^2=-1$.

(b) $(-1-i)^2+(-i)=(-1)^2+i+i+i^2+(-i)$

45. (a) First, note that $i(1+i)=-1+i$, $i^2(1+i)=-1-i$, $i^3(1+i)=-i+1$.
Plotting 1+*i* is just like plotting (1,1) in the Cartesian plane. Plotting -1+*i* is just like plotting (-1,1).

(b) $i(3-2i)=2+3i$, $i^2(3-2i)=-3+2i$, and $i^3(3-2i)=-2-3i$.

E. Mandelbrot Sequences

47. (a) $s_1=(-2)^2+(-2)$; $s_2=(s_1)^2+(-2)$; $s_3=(s_2)^2+(-2)$; $s_4=(s_3)^2+(-2)$.

(b) Strategy: Look for a pattern.
Don't worry, you should spot a pattern in (a) that you can apply.

51. (a) Parts (a), (b) and (c) of exercise 41 are particularly relevant.

(b) Strategy: Look for a pattern.

53. (a) Strategy: Use a variable.
This is an algebra problem at its heart. Since, $s_{N+1}=(s_N)^2+s$, you are being asked to solve $38=(6)^2+s$ for s.

(b) A pretty small value of *N* will work.

JOGGING

55. In step 1, three triangles are added. At each step thereafter, the number of triangles added (R) is quadrupled. The area S of each triangle added at a given step is 1/9 of the area added during the previous step. The total new area T added at a given step is the product of R and S. The area of the "snowflake" obtained at a given step (which we denote by Q) is the sum of T and the area Q of the snowflake in the previous step.

For step N, see the discussion found in the middle of page 442.

57. At the first step, a cube is removed from each of the six faces and from the center for a total of $C = 7$ cubes removed. At the next step, each of the 20 remaining cubes has 7 cubes removed (so $C = 20 \times 7$). In the second step, there are 20^2 remaining cubes each of which has 7 cubes removed.

63. Strategy: Use a variable.
Suppose that it is attracted to a number, say x. Then we have $s_{N+1} = (s_N)^2 + (0.25)$ and we set $x = s_{N+1} = s_N$. Substituting, we have $x = x^2 + 0.25$ so $x^2 - x + 0.25 = 0$.

65. Be bold. Collect several pieces of data by calculating several steps in the sequence. You may want to get your hands dirty on this one.

mini-excursion 3

A. Linear Growth and Arithmetic Sequences

1. (a) The common difference is what is added at each of the 30 stages of growth.

(b) If $P_0 = 75$ represents the original population, and $P_1 = 80$ represents the population after one generation, you are being been asked to find the first value of N so that $P_N \geq 1000$. That is, the first value of N such that $75 + 5 \cdot N \geq 1000$.

(c) Strategy: Work backwards.
You are being been asked to find the first value of N so that $P_N \geq 1002$.

3. (a) Strategy: Look for a pattern.
The common difference d is added at each of the 10 generations.

(b) The population grew by 30 after 10 generations. Since it grows linearly, it will grow by another 30 after 10 more generations. So, $P_{20} = 68$. Similarly, one can determine P_{30}, P_{40}, and P_{50}.

(c) Strategy: Look for a pattern.
You are being asked to express P_N in terms of N. N of the common differences d (found in (a)) are added to $P_0 = 8$ to find P_N.

5. (a) This sum is 2 plus how many 5's?

(b) First, determine the type of sequence that the terms of the sum form.

7. (a) Start by finding the common difference d. Then, determine how many of these are added to 12 to arrive at 309.

(b) Strategy: Look for a pattern.
Determine the type of sequence that appears in this problem. To find the sum, you will need to use the number of terms from part (a).

9. (a) Suppose that the number of streetlights at the end of week N is P_N.

(b) Find an explicit expression that represents P_N.

(c) The cost to operate one light for 52 weeks is \$52. To operate 137 lights is many times more expensive.

(d) Determine the yearly cost of operating a set of lights installed at the end of week N. Then, add these costs for N between 1 and 52.

B. Exponential Growth and Geometric Sequences

11. (a) The common ratio r is the value of $\frac{P_1}{P_0}$.

(b) P_9 can be found recursively from P_0 (by multiplying by the common ratio r over and over) or explicitly. The latter is the better choice here.

(c) Strategy: Look for a pattern.
You are being asked to express P_N in terms of *r* and P_0.

13. (a) You are being asked to express P_N in terms of P_{N-1}. But, $P_N = rP_{N-1}$ where $P_0 = 200$. Determine the value of *r* that one multiplies by each year to account for an *increase* by 50%.

(b) You are being asked to express P_N in terms of *N*.

(c) You are being asked to approximate the value of P_{10}. $P_{10} = 200 \times r^{10}$ and you need to find the value of the common ratio *r*.

C. Logistic Growth Model

15. (a) $p_1 = r \times (1 - p_0) \times p_0$

(b) $p_2 = r \times (1 - p_1) \times p_1$

(c) $p_3 = r \times (1 - p_2) \times p_2$; this decimal represents the percent of carrying capacity present in the third generation.

17. (b) Note that the values of p_N are getting very close to 0. This represents the percent of population present in generation *N*.

D. Miscellaneous

25. (a) There is a common ratio between terms.

(b) There is a common difference between terms.

(c) There is neither a common ratio nor a common difference between terms.

(d) There is a common ratio between terms. Can you find it?

(e) There is neither a common ratio nor a common difference between terms.

(f) There is a common difference between terms.

(g) There is a common difference (0) between terms. Then again, there is also a common ratio (1) between terms.

27. The first *N* terms of the arithmetic sequence are $c, c+d, c+2d, \ldots c+(N-1)d$.

29. Strategy: Work backwards.
This would require $p_{N+1} = p_N$. That is, $p_0 = p_1 = 0.8(1-p_0)p_0$.

Chapter 13

WALKING

A. Surveys and Public Opinion Polls

1. (a) The population in this study is the collection of objects that are under study.

(b) A sample is a subgroup of the population from which data is collected.

(c) What is the proportion of red gumballs in the sample?

(d) There are many different types of sampling methods. They include quota sampling, simple random sampling, convenience sampling, stratified sampling, and taking a census.

3. (a) A parameter is used to describe a population.

(b) A sample statistic is information derived from a sample. It may or may not reflect the actual population parameter. The difference in these values is attributed as sampling error.

(c) Flip a coin?

5. (a) Find the fraction of the population that makes up the sample.

(b) A sample statistic is information derived from a sample. It may or may not reflect the actual population parameter.

7. Sampling error is the difference between the population parameters (the actual vote) and the sample statistics.

9. (a) While this is a census, it is only one of a subset of the target population.

(b) What fraction of the target population gave their opinion in this survey?

11. (a) The population is the group that we are trying to get information about while the sampling frame consists only of those students that could have been surveyed using this method.

(b) The basic question is whether or not the sample chosen is representative of the population.

17. (a) Who is the survey trying to gain information about?

19. (a) Do you walk about your downtown in a random pattern? Do you think others do?

(b) Assume that people who live or work downtown are much more likely to answer yes than people in other parts of town.

(d) That is, was an attempt made to use quotas to get a representative cross section of the population?

21. (a) Does every member of the population have a chance of being in the sample using this survey technique?

23. (a) In simple random sampling, any two members of the population have as much chance of both being in the sample as any other two.

(b) The students sampled appear to be a reasonably fair cross section of all TSU undergraduates.

25. (a) The trees are broken into three different varieties and then a random sample is taken from each.

(b) When selecting 300 trees of variety A, the grower does not select them at random.

27. (a) George sure took the easy route to get survey results.

(b) Seems like a proportional sample.

(c) Seems pretty simple to me.

29. (a) Throw all the pineapples up in the air?

(b) Which pineapples are easiest to get at?

(c) There are three strata – one per supplier.

B. The Capture-Recapture Method

31. $\frac{n_1}{N} = \frac{n_2}{k}$

33. To estimate the number of quarters, remember to disregard the nickels and dimes—they are irrelevant.

37. Strategy: Use a variable.
Find n_2 ?

39. Strategy: Use logical reasoning.
You may first want to try an example with fictional numbers to see what would happen.

C. Clinical Studies

41. (a) The target population are those people that could benefit from the treatment. Who are they in this study?

(b) The sampling frame, in this case, is only a small portion of the target population. Be more specific in your response.

(c) Who would the sample likely under-represent?

47. (a) There was one group receiving the sham-surgery and two groups receiving a treatment.

(b) The treatments were specific types of knee surgery. Name them.

(c) A randomized study is one in which the patients are assigned to a treatment group or the control group at random.

(d) The question here boils down to who was it that knew the treatment being received by a given patient. Did the patient know? Did the doctors?

51. (b) All that is needed for an experiment is for members of the population to receive a treatment. It is a controlled placebo experiment if there is a control group that does not receive the treatment, but instead received a placebo. It is randomized if the participants are randomly divided into treatment and control groups. A study is double blind if neither the participants nor the doctors involved in the clinical trial know who is in each group.

57. (a) Use a complete sentence that is very detailed when answering questions such as this.

(b) The difference can sometimes be subtle. The sampling frame is those that could have been in the sample, but were not (for dumb luck in most cases). The target population is the group for which a result is being found.

D. Miscellaneous

61. (a) Was a treatment imposed on a subset of the population?

(b) What group of people is Spurlock trying to make a point about?

(c) What group of people received the treatment?

(d) There are plenty. Almost nothing in the study is well-done.

65. (a) This statement refers to the entire population of students taking the SAT math test.

(b) Only some new automobiles are crash tested.

(c) Not all of Mr. Johnson's blood was used for the test.

JOGGING

71. (a) In the real world, cost and reliability are important factors when making any decision.

(b) People with unlisted phone numbers make up a much higher percentage of the population in a large city such as New York. Also, what do you think is true about people with unlisted phone numbers?

73. Fridays, Saturdays, and Sundays make up 3/7 or about 43% of the week, but people put many more miles on their cars during the work week.

75. (a) How would *you* answer this survey (let's assume that you *do* cheat a bit on your taxes)?

Chapter 14

WALKING

A. Tables, Bar Graphs, Pie Charts, and Histograms

1. Strategy: Use a table or chart.
A frequency table will have the following format.

Score	10	50	…
Frequency	1	3	

3. (b) The bar graph will have five bars.

5. First, determine what percentage (or fraction) of children scored a 0.

7. Start by making a frequency table.

9. The central angle of the "very close" slice can be found from the frequency table. 7 of the 25 students fall in this category so that the central angle measures $\frac{7}{25}\times 360° = 100.8°$.

11. (a) four students scored a 3, six students scored a 4, etc…

13. (c) The sum of all of the central angles is 360 degrees.

17. Changing the scale of the vertical axis can really change how one interprets the graph.

19. (b) 16 oz. = 1 lb.
Also, remember that values that fall exactly on the boundary between two class intervals belong to the class interval to the left (due to the "or equal to" phrase).

(c) Remember that a histogram is slightly different than a bar graph.

B. Averages and Medians

23. (a) There isn't really a shortcut here – just grind it out.

(b) First, sort the data. Then, find the locator of the 50th percentile. Based on that, determine the value of the median.

(c) There are shortcuts to finding the average and the median of this new data set. Try to find them.

27. (a) Strategy: Look for a pattern. See chapter 1 for more on how to easily compute the sum $1 + 2 + 3 + \ldots + 98 + 99$. (Note: This is the sum of terms in an arithmetic sequence.)

(b) The locator for the median is $L = (0.50)(99) = 49.5$. Note: This locator is not a whole number.

29. (a) The data set consists of $24 + 16 + 20 + \ldots + 3$ values. 24 of them are 0, 16 are 1, 20 are 2, etc.

(b) The locator for the median is $L = (0.50)(80) = 40$. So, the median is the average of the 40th and 41st scores. Use the table to determine the 40th and 41st scores (in the order data set).

31. (a) The number of scores is not important. If one likes, they can assume that there were 100 scores.

(b) Half of the scores are at or below what score?

C. Percentiles and Quartiles

33. (a) The locator of the 25th percentile is $L = (0.25)(9) = 2.25$. So, the first quartile is the 3rd number in the ordered list.

(b) The locator of the 75th percentile is $L = (0.75)(9) = 6.75$.

(c) The locator of the 25th percentile of this new data set is $L = (0.25)(10) = 2.5$. The first quartile is still the 3rd number in the ordered list.

37. (a) The Cleansburg Fire Department consists of $N = 2 + 7 + 6 + \ldots$ firemen. The locator of the first quartile is thus given by $L = (0.25)(N)$.

D. Box Plots and Five-Number Summaries

41. (a) The five number summary consists of these five numbers: minimum, first quartile, median, third quartile, maximum.

(b) Strategy: Draw a diagram. First, draw and label an appropriate axis. One such grid is shown below.

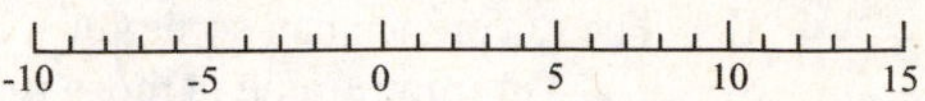

45. (a) One of the vertical line segments in the box plot will tell you this information.

(c) Two different vertical line segments in these box plots should explain why this happens.

E. Ranges and Interquartile Ranges

47. (a) The range is a single number. Your (common sense) answer should be too.

(b) The IQR is also one number.

51. (a) Note that $1.5 \times IQR = 1.5 \times 3 = 4.5$. So any number more than this amount to the *right* of a box in a box plot is an outlier.

(b) Any number more than $1.5 \times IQR$ to the *left* of the box in a box plot is an outlier.

F. Standard Deviations

55. (a) Just how much does this data set deviate from the mean? Do this by inspection if you can.

(b) The average of the data set is 5. So, the following table is useful. The final step will be to take the square root of the average of the squared deviations.

x	$x-5$	$(x-5)^2$
0	-5	25
5	0	0
5	0	0
10	5	25
		50

57. There is a lesson to be learned in doing each of these parts. Standard deviation measures the spread of a data set. Some of these data sets are spread out in the exact same fashion. For part (a), fill out the table shown below:

Data	*Deviation from mean*	*Squared Deviations from mean*
0	–4.5	20.25
1	–3.5	12.25
2	–2.5	6.25
3	–1.5	2.25
4	–0.5	0.25
5	0.5	0.25
…	…	…

G. Miscellaneous

59. One easy way to remember the mode: MOde = MOst common.

JOGGING

65. Strategy: Use a variable.
Let x = the score Mike needs on the next exam. Find the value of x.

67. Strategy: Use a variable.
$(0.5)(N) = L$. You are basically told the value of L. So find N.

71. Figure (a) shows a median between \$70 and \$100 million.

77. Compute

$$\frac{(x_1+c)+(x_2+c)+(x_3+c)+\ldots+(x_N+c)}{N}.$$

Group the N c's together.

79. Range is a measure of spread. Plot an example on a number line (take any data set and, say, c = 2). This will help you "see" the reason.

Chapter 15

WALKING

A. Random Experiments and Sample Spaces

1. (a) The sample space of a random experiment is the set of all its possible outcomes. One outcome, for example, might be denoted *HTT* (heads on the first toss, tails on the second toss, heads on the third toss).

(b) You may want to think of success *s* in as "Heads" and failure *f* as "Tails" in (a).

(c) The coin may never land heads. Then again, it could always land heads. The sample space will contain each possible number of heads that could be tossed.

(d) A free throw is either successful or not. It is pretty black and white. How many free throws could be made in three tosses?

3. {*ABCD*, *ABDC*, *ACBD*, …}
(list 'em all)

5. A typical outcome is a string of 10 letters each of which can be either an *H* or a *T*. Be as descriptive as you can. There are a lot of outcomes - two possible outcomes at each of 10 stages of the experiment.

B. The Multiplication Rule

9. (a) You could think of constructing a license plate as a seven stage process. There are 9 possibilities in the first stage, 26 in the second, 26 in the third, etc. [Also, remember that you have 10 digits…they are called fingers…you could label them as 0-9.]

(b) Seems like only one choice at the first and last (seventh) stage of constructing such a plate.

(c) There are 9 possibilities in the first stage, 26 in the second, 25 in the third (no repeats), etc.

11. (a) Dolores could either wear
(1) high heels and a formal dress or
(2) high heels, jeans, and a blouse or
(3) tennis shoes, jeans, and a t-shirt

There are $2\times4=8$ outfits of type (1) that she could wear.

(b) In addition to the options in (a), there is a fifth category: (5) tennis shoes, jeans, t-shirt, and sweater.

13. (a) 8 choices for the label of the book at the far left, 7 choices for the label of the book to its right, 6 choices for the label of the book to its right, etc.

(b) How many ways for the books to be in order?

15. (a) Select the committee in three stages. First, select a President (35 choices), then, select a VP (34 choices), etc.

(b) Select the committee in three stages. First, select a President from the eligible members, then, select a VP from the remaining eligible members, etc.

(c) How many all-female committees are possible?
How many all-male committees are possible?

(d) In all possible committees (there are ${}_{35}P_3$ of them), the members are either all of the same gender or not.

17. (a) The only restrictions seem to be that the first digit is not zero and that the last digit is even.

(b) Do you remember your rules of divisibility?

(c) Strategy: Look for a pattern.
List several numbers that are divisible by 25 if you need to in order to spot a pattern.

C. Permutations and Combinations

19. (a) ${}_nP_r=\dfrac{n!}{(n-r)!}$

(b) ${}_nC_r=\dfrac{n!}{r!(n-r)!}$

(c) Strategy: Look for a pattern.

$${}_{10}P_3=\frac{10!}{7!}=\frac{10\times9\times8\times\cdots\times2\times1}{7\times6\times5\times\cdots\times2\times1}.$$

Can you simplify this expression before performing any multiplication?

23. ${}_{11}C_4=\dfrac{11!}{4!7!}={}_{11}C_7$

25. The point of this exercise is to illustrate a nice pattern.

27. ${}_{150}P_{51}=\dfrac{150\times149\times148\times\cdots\times3\times2\times1}{99\times98\times97\times\cdots\times3\times2\times1}$

Write out the value of ${}_{150}P_{50}$ in the same way to determine how many times larger than ${}_{150}P_{50}$ this is.

29. Strategy: Solve a simpler problem first. Argue in the same way you did in exercise 27.

31. Sometimes the ${}_nP_r$ and ${}_nC_r$ keys can be found under a menu (such as MATH or PRB) on the calculator. Also, remember that the two numbers appearing on the right of E represent a power of 10 in scientific notation.

33. (a) Are the selections ordered or unordered? That is the question!

(b) That is always (well often at least) the question!

35. (a) The order that the songs are selected to be played on tour do not matter – apparently all that is relevant here is *which* songs are played.

(b) In the old days, songs were put on records or cassettes. Since it was much more difficult to change songs while listening, the sequencing of songs was critical. The Beatles *White Album* is a prime example of how critical order is to an album.

The arrangement of the songs on The White Album follows patterns and establishes symmetries that have been much analyzed over the years. For example, "Wild Honey Pie" is the fifth song from the beginning of the album and "Honey Pie" is the fifth song from the end. Also, three of the four songs containing animal names in their titles ("Blackbird", "Piggies", and "Rocky Raccoon") are grouped together. In a similar fashion, "Honey Pie" and "Savoy Truffle" (both referring to types of desserts) play back to back towards the end of the album.

"Savoy Truffle", the fourth song from the end, contains a reference to "Ob-La-Di, Ob-La-Da," the fourth song from the beginning. In addition, the album's four Harrison compositions are distributed evenly, with one appearing on each of the four sides. Each LP's first track is a McCartney composition marking a return to traditional rock n' roll ("Back in the USSR" and "Birthday"). Each LP concludes with a Lennon composition built around themes of childhood and innocence. ("Julia" and "Good Night.") Even the number of tracks per side correspond to the number of letters in each of the Beatles' names, with Side 1 for Harrison, Side 2 for McCartney, Side 3 for Starkey, and Side 4 for Lennon.[1]

D. General Probability Spaces

37. (a) Strategy: Use a variable.
The probabilities of the outcomes must sum to 1 in order to have a probability space. Let $x = \Pr(o_3) = \Pr(o_4) = \Pr(o_5)$.

(b) $\Pr(o_3) = \Pr(o_4) + \Pr(o_5) = \Pr(o_4) + 0.1$

39. Strategy: Use a variable.
Let $x = \Pr(B)$. Then, $\Pr(C) = 2x$ and $\Pr(D) = 3x$. Note that the sum of the probabilities must equal 1.

41. Not all of the outcomes are equally likely. The probability of landing in a particular sector is proportional to the size of that sector.

E. Events

43. (a) Strategy: Make an organized list.
$E_1 = \{HHT, ...\}$

(b) E_2 consists of two outcomes. List them.

(c) E_3 doesn't have much in it. But it can still be written using set notation. Remember that an event is a *set* of outcomes.

47. (b) Strategy: Make an organized list.
The last toss could be tails or the second to last toss could be tails or the third to last toss could be tails or…

(c) This means that either exactly 9 or exactly 10 tosses come up tails.

F. Equiprobable Spaces

49. Exercise 43 should help you with the numerators. The size of the entire (equiprobable) sample space will help you with the denominator.

53. Again, exercise 47 should help you with the numerators. The size of the entire (equiprobable) sample space will help you with the denominator.

55. You have two choices on the first problem (*T* or *F*), two choices on the second problem, etc. The total number of outcomes in this random experiment (the denominator in these problems) is found by the multiplication rule.

(a) There is only one way to get all ten correct. The key must match the answer sheet. Compare this situation to exercise 53(a).

(b) There is only one way to get all ten incorrect (and hence a score of –5). The key must completely disagree with the answer sheet. See also exercise 53(a).

(c) In order to get 8.5 points, the student must get exactly 9 correct answers and 1 incorrect answer. Any one of 10 different questions could be answered incorrectly. See also exercise 53(b).

(d) In order to get 8 or more points, the student must get at least 9 correct answers (if they only get 8 correct answers, they lose a point for guessing two incorrect answers and score 7 points).

(e) If the student gets 7 answers correct, they score $7 - 3 \times 0.5 = 5.5$ points. How can they score exactly 5 points?

(f) If the student gets 8 answers correct, they score $8 - 2 \times 0.5 = 7$ points. How many ways are there to select which 8 answers would be answered correctly (or, for that matter, which two would be answered incorrectly)?

57. (a) How many ways are there to choose four delegates? Assuming Alice is chosen,

[1] "The Beatles (album)." *Wikipedia, The Free Encyclopedia*. 5 Oct 2008 <http://en.wikipedia.org/w/index.php?title=The_Beatles_(album)&oldid=243125771>.

how many ways are there to choose the other three delegates?

(b) Part (a) will come in handy here.

G. Odds

59. **(a)** If the probability of event E is a/b than the odds in favor of E are a to b-a.

(b) Since 0.6 is rational, it can be written as a fraction.

61. If the odds in favor of event E are m to n than the probability of E is $m/(m+n)$.

JOGGING

63. Select the chair first and then the secretary. At the next stage, select the other three members.

65. **(a)** Strategy: Make an organized list. The outcomes in this event must be "of length 5" and "end in X winning."

(b) Strategy: Make an organized list. The outcomes in this event must be "of length 5."

(d) First, count all of the outcomes that consist of X winning the series. X must be listed last in such an outcome. So, you want to count the number of outcomes of length 4, 5, 6, and 7 in which X is listed last and X appears three times before that. Double this number to find your answer.

67. **(b)** Strategy: Solve a simpler problem first. A circle of 10 people can be broken to form a line in 10 different ways. Try this problem with 2, 3, or 4 people first.

(c) There are two choices as to which sex will start the line. Then, count how many ways there are to order the boys (the next stage) and how many ways there are to order the girls in the line (the last stage).

69. Form the groups in three stages.

71. **(a)** The answer is *not* ½.

(b) It will be useful to count how many ways one can choose the positions of the H's.

(c) Pr(3 or more H's) = 1 – Pr(0 H's) – Pr(1 H's) – Pr(2 H's)

73. The total number of 5-card *draw* poker hands (see Example 15.13) is 2,598,960. How many hands have all 5 cards the same color? [The first card can be any color, the remaining cards must match that color.]

75. Count the number of ways to get 10, J, Q, K, A of any suit (including all the same suit) and subtract the number of ways for these cards to all be the same suit.

77. $\Pr(\text{win Bet 1}) = 1 - \Pr(\text{never roll a 6})$;

$\Pr(\text{win Bet 2}) = 1 - \Pr(\text{never roll boxcars})$

79. Pay attention to the phrase "at least." You could win either 1 prize, 2 prizes, or 3 prizes. Add the probability of each event happening.

Chapter 16

WALKING

A. Normal Curves

1. **(a)** Remember that in a normal distribution, the mean and the median are located at the center of the distribution.

(c) In a normal distribution, the standard deviation equals the horizontal distance between a point of inflection and the (vertical) axis of symmetry for the distribution.

3. **(a)** See the hint for 1(a).

(b) See the hint for 1(c).

(c) $Q_3 = \mu + 0.675\sigma$

(d) $Q_1 = \mu - 0.675\sigma$

7. **(a)** Strategy: Look for symmetry.

(b) Use the fact that $Q_3 = \mu + 0.675\sigma$ and solve for σ.

9. See the hint to Exercise 7(b).

11. See the hint for Exercise 1(a).

13. Normal distributions are very symmetric.

B. Standardizing Data

15. (a) For a normal distribution with mean μ and standard deviation σ, a data value x has a standardized value z obtained by subtracting the mean μ from x and dividing the result by the standard deviation σ. In other words, $z = \frac{x-\mu}{\sigma}$.

(b) This can be done by inspection. Also, since the units of $x-\mu$ (kg) and the units of σ (kg) are the same, the value of z has no units.

17. (a) Use $Q_3 \approx \mu + (0.675)\sigma$ to first determine the value of σ. Then use the fact that $z = \frac{x-\mu}{\sigma}$.

21. Solving $z = \frac{x-\mu}{\sigma}$ for x gives $x = \mu + \sigma \cdot z$.

23. Solve $z = \frac{x-\mu}{\sigma}$ for σ.

25. Solve $z = \frac{x-\mu}{\sigma}$ for μ.

27. Solve for μ and for σ:

$$\frac{20-\mu}{\sigma} = -2; \frac{100-\mu}{\sigma} = 3$$

C. The 68 – 95 – 99.7 Rule

29. In every normal distribution, approximately 68% of all the data values fall within 1 standard deviation above and below the mean. Also, 95% of all the data values fall within 2 standard deviations above and below the mean. Approximately 99.7% of all the data values fall within 3 standard deviations above and below the mean. Moreover, 50% of the data lies between the quartiles.

31. See the hint for exercise 29.

35. In a normal distribution, approximately 84% of the data lies above one standard deviation of the mean.

37. (a) Since 95% of the data lies within two standard deviations of the mean, 2.5% of the data are not within two standard deviations on each side of the mean.

(b) This is half of the amount of data that falls between 1 and 2 standard deviations of the mean.

39. Convert these values to standardized values in order to determine their percentiles.

D. Approximately Normal Data Sets

41. (c) 41 has a standardized value of -1 and 63 has a standardized value of 1.

(d) 63 is one standard deviation above the mean. Also, 68% of the data is within one standard deviation of the mean (so that 32% of the data is either more than one standard deviation above the mean or less than one standard deviation below the mean).

45. (a) 99 is two standard deviations below the mean; 151 is two standard deviations above the mean. Use the 68-95-99.7 rule.

(b) 112 is one standard deviation below the mean. 151 is two standard deviations above the mean.

47. A value two standard deviations below the mean is at about the 3rd percentile. A value one standard deviation below the mean is at about the 16th percentile. A value one standard deviation above the mean is at about the 84th percentile and a value two standard deviations above the mean is at about the 97th percentile.

53. (a) 15.25 has a standardized value of –1.

(b) 21.25 has a standardized value of 2.

(c) The 75th percentile is 0.675 standard deviations above the mean.

E. The Honest and Dishonest Coin Principles

57. (a) According to the honest-coin principle, the random variable Y is approximately normal with mean $\mu = \frac{n}{2}$ and standard deviation $\sigma = \frac{\sqrt{n}}{2}$.

(b) 1770 is approximately one standard deviation below the mean. 1830 is approximately one standard deviation above the mean.

(d) 1860 is approximately two standard deviations above the mean.

59. According to the honest-coin principle, the number of females in the sample is approximately normal with mean $\mu = \frac{n}{2}$ and standard deviation $\sigma = \frac{\sqrt{n}}{2}$. Convert all the given numerical values to standardized values.

61. If p is the probability of the coin landing heads, then X has an approximately normal distribution with mean $\mu = np$ and standard deviation $\sigma = \sqrt{np(1-p)}$.

63. The die is behaving like a dishonest coin with $p = 1/6$. There are two possible outcomes (a "6" and not a "6").

65. The numbers 117 and 139 are each two standard deviations away from the mean. The mean tells the probability of a widget selected at random being defective.

JOGGING

67. **(a)** According to the table, the 95th percentile has a value of $\mu + 1.65\sigma$.

(b) According to the table, the 40th percentile has a value of $\mu - 0.25\sigma$.

69. Work in terms of standardized values.

73. **(a)** According to the table, the 90th percentile of the data is located at $\mu + 1.28 \times \sigma$ points.

(b) The 70th percentile of the data is located at $\mu + 0.52 \times \sigma$ points.

75. In a normal distribution, 16% of the data is more than one standard deviation above the mean $\mu = \frac{n}{2}$. Also, by the honest-coin principle, $\sigma = \frac{\sqrt{n}}{2}$.

mini-excursion 4

A. Weighted Averages

1. Paul's score is a weighted average $0.15 \times 77\% + 0.15 \times 83\% + \ldots =$

B. Expected Values

5. Look back at the definition of expected value found on page 616.

7. **(a)** A probability distribution will look like the table found in exercises 5 and 6.

(c) A "fair game" is one where the expected outcome is 0.

9. **(a)** First, write out a sample space in which all the outcomes are equiprobable.

11. **(a)** The probability of winning \$1 is 18/38. The probability of losing \$1 (that is, gaining -\$1) is 20/38.

13. **(b)** Use the expected value found in (a).

C. Miscellaneous

15.

Outcome (Benefit to Joe)
Probability

Fill in the table above and use it to compute an expected value.

17.

Payoff (to insurer)	$\$P$			
Probability	50%	35%	12%	3%

Fill in the table above. Compute the expected value in terms of the policy price $\$P$. Set the expected value equal to \$50 and solve for $\$P$.

19. **(a)** Use the multiplication rule.

Chapter 1

WALKING

A. Ballots and Preference Schedules

1. (a)

Number of voters	5	3	5	3	2	3
1st choice	A	A	C	D	D	B
2nd choice	B	D	E	C	C	E
3rd choice	C	B	D	B	B	A
4th choice	D	C	A	E	A	C
5th choice	E	E	B	A	E	D

(b) A has 5 + 3 = 8 first-place votes, B has 3 first-place votes, C has 5 first-place votes, and D has 3 + 2 = 5 first-place votes. So A has a plurality (the most) of first-place votes.

(c) A has 3 last-place votes, B has 5 last-place votes, C has no last-place votes, D has 3 last-place votes and E has 5 + 3 + 2 = 10 last-place votes. So C has the fewest last-place votes (with none).

3. (a) 5 + 3 + 5 + 3 + 2 + 3 = 21

(b) There are 21 votes all together. A majority is more than half of the votes, or at least 11.

(c) Argand is favored over Dietz by 5 + 3 + 3 = 11 (a majority) of the voters.

5. (a) No candidate has a majority of the first-place votes. So, all candidates with 20% or less of the 21 first-place votes are eliminated. Since 0.20(21) = 4.2, all candidates with fewer than 4.2 first-place votes (4 votes or fewer) are eliminated. The candidates that are eliminated are B (3 first-place votes) and E (0 first-place votes).

(b)

Number of voters	5	3	5	3	2	3
1st choice	A	A	C	D	D	A
2nd choice	C	D	D	C	C	C
3rd choice	D	C	A	A	A	D

This schedule can be more efficiently written as follows:

Number of voters	8	3	5	5
1st choice	A	A	C	D
2nd choice	C	D	D	C
3rd choice	D	C	A	A

(c) Professor Argand now has 11 first-place votes and is the majority winner.

7.

Number of voters	255	480	765
1st choice	L	C	M
2nd choice	M	M	L
3rd choice	C	L	C

(0.17)(1500) = 255; (0.32)(500) = 480; The remaining voters (51% or 1500-255-480=765) prefer M the most, C the least, so that L is their second choice.

9.

Number of voters	47	36	24	13	5
1st choice	*B*	*A*	*B*	*E*	*C*
2nd choice	*E*	*B*	*A*	*B*	*E*
3rd choice	*A*	*D*	*D*	*C*	*A*
4th choice	*C*	*C*	*E*	*A*	*D*
5th choice	*D*	*E*	*C*	*D*	*B*

Think of the numbers in each column as representing the position of the corresponding letter (candidate) from top to bottom in that column.

B. Plurality Method

11. (a) *A* has 153 + 102 + 55 = 310 first-place votes.
B has 202 + 108 + 20 = 330 first-place votes.
C has 110 + 160 = 270 first-place votes.
D has 175 + 155 = 330 first-place votes.
B and *D* tie with 330 first-place votes each.

(b) *B* has 55 + 110 + 175 = 340 last-place votes.
D has 153 + 20 + 160 = 333 last-place votes.
With the fewest last-place votes, *D* wins the election.

(c) *B* is preferred to *D* by 153 + 102 + 202 + 108 + 20 + 160 = 745 of the voters.
D is preferred to *B* by 55 + 110 + 175 + 155 = 495 of the voters.
B wins the election.

13. (a) 23 votes will guarantee *A* at least a tie for first; 24 votes guarantee that *A* is the only winner. (With 23 of the remaining 30 votes *A* has 49 votes. The only candidate with a chance to have that many votes is *C*. Even if *C* gets the other 7 remaining votes, *C* would not have enough votes to beat *A*.)

For another way to compute this answer, suppose that *A* receives *x* out of the remaining 30 votes and the competitor with the most votes, *C*, receives the remaining 30 - *x* votes. To determine the values of *x* that guarantee a win for *A*, we solve $26+x \geq 42+(30-x)$, i.e. $2x \geq 46$. So $x \geq 23$ votes will guarantee *A* at least a tie for first.

(b) 11 votes will guarantee *C* at least a tie for first; 12 votes guarantee *C* is the only winner. (With 11 of the remaining votes *C* has 53 votes. The only candidate with a chance to have that many votes is *D*. Even if *D* gets the other 19 remaining votes, *D* would not have enough votes to beat *C*.)

For another way to compute this answer, suppose that *C* receives *y* out of the remaining 30 votes and the competitor with the most votes, *D*, receives the remaining 30 - *y* votes. To determine the values of *y* that guarantee a win for *C*, we solve $42+y \geq 34+(30-y)$, i.e. $2y \geq 22$. So $y \geq 11$ votes will guarantee *C* at least a tie for first.

15. (a) 721/5 = 144.2 so that 145 votes are needed to have a plurality. In that case, each of the other candidates would have 144 votes.

(b) 10; If there were only 9 candidates in the election, then 81 votes would be needed for a plurality since $721/9 \approx 80.1$. With 10 candidates, 73 votes would produce a plurality since $721/10 \approx 72.1$. 73 votes would not be necessary for a plurality with 11 candidates ($721/11 \approx 65.5$ so that only 66 votes would be needed in this case).

(c) 7; If there were only 6 candidates in the election, then 121 votes would be needed for a plurality since $721/6 \approx 120.2$. With 7 candidates, 104 votes would produce a plurality since $721/7 = 103$.

104 votes would not be necessary for a plurality with 8 candidates ($721/8 \approx 90.1$ so that only 91 votes would be needed in this case).

C. Borda Count Method

17. **(a)** A has $5\times(5+3)+4\times0+3\times3+2\times(5+2)+1\times3=66$ points.

B has $5\times3+4\times5+3\times(3+3+2)+2\times0+1\times5=64$ points.

C has $5\times5+4\times(3+2)+3\times5+2\times(3+3)+1\times0=72$ points.

D has $5\times(3+2)+4\times3+3\times5+2\times5+1\times3=65$ points.

E has $5\times0+4\times(5+3)+3\times0+2\times3+1\times(5+3+2)=48$ points.

The winner is Professor Chavez.

(b)

Number of voters	5	3	5	5	3
1st choice	*A*	*A*	*C*	*D*	*B*
2nd choice	*B*	*D*	*D*	*C*	*A*
3rd choice	*C*	*B*	*A*	*B*	*C*
4th choice	*D*	*C*	*B*	*A*	*D*

(c) A has $4\times(5+3)+3\times3+2\times5+1\times5=56$ points.

B has $4\times3+3\times5+2\times(3+5)+1\times5=48$ points.

C has $4\times5+3\times5+2\times(5+3)+1\times3=54$ points.

D has $4\times5+3\times(3+5)+2\times0+1\times(5+3)=52$ points.

The winner is Professor Argand.

(d) Professor Chavez (choice *C*) was the winner of an election and when an irrelevant alternative (candidate *E*) was disqualified and the ballots recounted, we found that Professor Argand (choice *A*) was the winner of the reelection. This is a violation of the Independence-of-Irrelevant-Alternatives Criterion (IIA).

19. **(a)** A has $5\times8+4\times(2+1)+3\times7+2\times0+1\times6=79$ points.

B has $5\times0+4\times(8+7+6)+3\times2+2\times1+1\times0=92$ points.

C has $5\times2+4\times0+3\times8+2\times(7+6)+1\times1=61$ points.

D has $5\times(7+6)+4\times0+3\times1+2\times(8+2)+1\times0=88$ points.

E has $5\times1+4\times0+3\times6+2\times0+1\times(8+7+2)=40$ points.

The winner is Borrelli.

(b) Dante has a majority of the first-place votes (13 out of a possible 24 votes) but does not win the election.

(c) Dante, having a majority of the first-place votes, is a Condorcet candidate but does not win the election.

21. Assume, for the sake of simplicity, that there are $20N$ voters.
Column 1: $0.40\times20N=8N$ voters; Column 2: $0.25\times20N=5N$ voters;
Column 3: $0.20\times20N=4N$ voters; Column 4: $0.15\times20N=3N$ voters.
A has $4\times8N+3\times3N+2\times4N+1\times5N=54N$ points.
B has $4\times(4N+3N)+3\times5N+2\times8N+1\times0N=59N$ points.
C has $4\times5N+3\times0N+2\times0N+1\times(8N+4N+3N)=35N$ points.
D has $4\times0N+3\times(8N+4N)+2\times(5N+3N)+1\times0N=52N$ points.
Regardless of the value of *N*, the winner is *B*.

23. **(a)** To achieve the maximum number of points possible, a candidate would need to receive all 110 first-place votes. Such a candidate would have $4\times110+3\times0+2\times0+1\times0=440$ points.

(b) To achieve the minimum number of points possible, a candidate would need to receive all 110 last-place votes. Such a candidate would have $4\times0+3\times0+2\times0+1\times110=110$ points.

(c) 4 + 3 + 2 + 1 = 10 points

(d) $\frac{10\text{ points}}{1\text{ ballot}}\times110\text{ ballots}=1100\text{ points}$

(e) $1100-320-290-180=310$ points

25. *C* wins with 500 – 115 – 120 – 125 = 140 points; the points sum to $4\times50+3\times50+2\times50+1\times50=500$.

D. Plurality-with-Elimination Method

27. **(a)** Professor Argand
Round 1:

Candidate	*A*	*B*	*C*	*D*	*E*
Number of first-place votes	8	3	5	5	0

E is eliminated.
Round 2: Eliminating *E* does not affect the first-place votes of the candidates. Using the table from Round 1, we see that *B* now has the fewest number of first-place votes, so *B* is eliminated.
Round 3: The three votes originally going to *B* would next go to *E*, but *E* has also been eliminated. So *B*'s three votes go to *A*.

Candidate	*A*	*B*	*C*	*D*	*E*
Number of first-place votes	11		5	5	

Candidate *A* now has a majority of the first-place votes and is declared the winner.

(b) Since Professor Chavez is not under consideration, the resulting preference schedule would contain only four candidates.

Number of voters	5	3	5	3	2	3
1st choice	*A*	*A*	*E*	*D*	*D*	*B*
2nd choice	*B*	*D*	*D*	*B*	*B*	*E*
3rd choice	*D*	*B*	*A*	*E*	*A*	*A*
4th choice	*E*	*E*	*B*	*A*	*E*	*D*

(c) Professor Epstein.
Round 1:

Candidate	*A*	*B*	*D*	*E*
Number of first-place votes	8	3	5	5

B is eliminated.
Round 2: The three votes originally going to *B* go to *E*.

Candidate	*A*	*B*	*D*	*E*
Number of first-place votes	8		5	8

D is eliminated.
Round 3: The 5 votes that *D* had in round 2 would go to *B* except that *B* has been eliminated. Instead, three of these votes go to *E* and two of these five votes go to *A*.

Candidate	*A*	*B*	*D*	*E*
Number of first-place votes	10			11

Candidate *E* now has a majority of the first-place votes and is declared the winner.

(d) A was the winner of the original election, and, in the reelection without candidate C (an irrelevant alternative) it turned out that candidate E was declared the winner. This violates the Independence-of-Irrelevant-Alternatives criterion.

29. (a) Dante is the winner.
Round 1:

Candidate	A	B	C	D	E
Number of first-place votes	8	0	2	13	1

Candidate D has a majority of the first-place votes and is declared the winner.

(b) Since Dante is a majority winner, in this case the winner is determined in the first round.

(c) If there is a choice that has a majority of the first-place votes, then that candidate will be the winner under the plurality-with-elimination method. So, the plurality-with-elimination method satisfies the majority criterion.

31. B is the winner.
Round 1:

Candidate	A	B	C	D
Percent of first-place votes	40	35	25	0

D is eliminated.
Round 2: Eliminating D does not affect the first-place votes of the other candidates. Using the table from Round 1, we see that candidate C now has the smallest percentage of first-place votes, so C is eliminated.
Round 3: The 25% of the votes originally going to C now go to B.

Candidate	A	B	C	D
Percent of first-place votes	40	60		

Candidate B now has a majority of the first-place votes and is declared the winner.

33. (a) Dungeons (South Africa) is the winner. Round 1:

Candidate	A	B	C	D
Number of first-place votes	10	9	5	7

Candidate C is eliminated.
Round 2: The five votes originally going to C now go to D.

Candidate	A	B	C	D
Number of first-place votes	10	9		12

Candidate B is eliminated.
Round 3: Of the 9 votes going to B in Round 2, four will go to D and five will go to A.

Candidate	A	B	C	D
Number of first-place votes	15			16

Candidate D has a majority of the first-place votes and is declared the winner.

(b) B is the Condorcet candidate.
In a head-to-head contest, B beats A, 16 votes to 15 votes.
In a head-to-head contest, B beats C, 16 votes to 15 votes.
In a head-to-head contest, B beats D, 19 votes to 8 votes.

(c) Banzai Pipeline (B), which is the Condorcet candidate, fails to win the election under the plurality-with-elimination method. So, the plurality-with-elimination method fails to satisfy the Condorcet criterion.

E. Pairwise Comparisons Method

35. (a) *C* is the winner.

A versus *B*: 13 votes to 13 votes (tie). *A* gets $\frac{1}{2}$ point, *B* gets $\frac{1}{2}$ point.
A versus *C*: 8 votes to 18 votes (*C* wins). *C* gets 1 point.
A versus *D*: 14 votes to 12 votes (*A* wins). *A* gets 1 point.
A versus *E*: 21 votes to 5 votes (*A* wins). *A* gets 1 point.
B versus *C*: 8 votes to 18 votes (*C* wins). *C* gets 1 point.
B versus *D*: 8 votes to 18 votes (*D* wins). *D* gets 1 point.
B versus *E*: 8 votes to 18 votes (*E* wins). *E* gets 1 point.
C versus *D*: 13 votes to 13 votes (tie). *C* gets $\frac{1}{2}$ point, *D* gets $\frac{1}{2}$ point.
C versus *E*: 13 votes to 13 votes (tie). *C* gets $\frac{1}{2}$ point, *E* gets $\frac{1}{2}$ point.
D versus *E*: 15 votes to 11 votes. *D* gets 1 point.

The final tally, is $2\frac{1}{2}$ points for *A*, $\frac{1}{2}$ point for *B*, 3 points for *C*, $2\frac{1}{2}$ points for *D*, and $1\frac{1}{2}$ points for *E*. Candidate *C* is the winner.

(b) The preference schedule without Alberto is shown below.

Number of voters	8	6	5	5	2
1st choice	*C*	*E*	*E*	*D*	*D*
2nd choice	*B*	*D*	*C*	*C*	*E*
3rd choice	*D*	*B*	*D*	*E*	*B*
4th choice	*E*	*C*	*B*	*B*	*C*

(c) Without candidate *A*, the method of pairwise comparisons produces the following results.
B versus *C*: 8 votes to 18 votes (*C* wins). *C* gets 1 point.
B versus *D*: 8 votes to 18 votes (*D* wins). *D* gets 1 point.
B versus *E*: 8 votes to 18 votes (*E* wins). *E* gets 1 point.
C versus *D*: 13 votes to 13 votes (tie). *C* gets $\frac{1}{2}$ point, *D* gets $\frac{1}{2}$ point.
C versus *E*: 13 votes to 13 votes (tie). *C* gets $\frac{1}{2}$ point, *E* gets $\frac{1}{2}$ point.
D versus *E*: 15 votes to 11 votes. *D* gets 1 point.

The final tally, is 0 points for *B*, 2 points for *C*, $2\frac{1}{2}$ points for *D*, and $1\frac{1}{2}$ points for *E*. Candidate *D*, Dora, is now the winner.

(d) *C* was the winner of the original election, and, in the re-election without candidate *A* (an irrelevant alternative) it turned out that candidate *D* was declared the winner. This violates the Independence-of-Irrelevant-Alternatives criterion.

37. *A* versus *B*: 13 votes to 8 votes (*A* wins). *A* gets 1 point.
A versus *C*: 11 votes to 10 votes (*A* wins). *A* gets 1 point.
A versus *D*: 11 votes to 10 votes (*A* wins). *A* gets 1 point.
A versus *E*: 10 votes to 11 votes (*E* wins). *E* gets 1 point.
B versus *C*: 11 votes to 10 votes (*B* wins). *B* gets 1 point.
B versus *D*: 8 votes to 13 votes (*D* wins). *D* gets 1 point.
B versus *E*: 16 votes to 5 votes (*B* wins). *B* gets 1 point.
C versus *D*: 13 votes to 8 votes (*C* wins). *C* gets 1 point.
C versus *E*: 18 votes to 3 votes (*C* wins). *C* gets 1 point.

D versus *E*: 13 votes to 8 votes (*D* wins). *D* gets 1 point.
The final tally is 3 points for *A*, 2 points for *B*, 2 points for *C*, 2 points for *D*, and 1 point for *E*. Candidate *A*, Professor Argand, is the winner.

39. (a) With five candidates, there are a total of 4 + 3 + 2 + 1 = 10 pairwise comparisons. Each candidate is part of four of these. So, to find the number of points each candidate earns, we simply subtract the losses from 4. The 10 points are distributed as follows: *E* wins $1\frac{1}{2}$ points, *D* wins $2\frac{1}{2}$ points, *C* gets 3 points, *B* gets 2 points, and *A* gets the remaining $10-1\frac{1}{2}-2\frac{1}{2}-3-2=1$ point.

(b) Candidate *C*, with 3 points, is the winner.

F. Ranking Methods

41. (a) *A* has 10 first-place votes.
B has 5 + 4 = 9 first-place votes.
C has 5 first-place votes.
D has 7 first-place votes.
Winner *A*: Second place: *B*. Third place: *D*. Last place: *C*.

(b) *A* has $4\times10+3\times0+2\times(7+5+5)+1\times4=78$ points.
B has $4\times(5+4)+3\times7+2\times10+1\times5=82$ points.
C has $4\times5+3\times(10+5+4)+2\times0+1\times7=84$ points.
D has $4\times7+3\times5+2\times4+1\times(10+5)=66$ points.
Winner: *C*. Second place: *B*. Third place: *A*. Last place: *D*.

(c) Round 1:

Candidate	*A*	*B*	*C*	*D*
Number of votes	10	9	5	7

C is eliminated.
Round 2: The five votes originally going to *C* now go to *D*.

Candidate	*A*	*B*	*C*	*D*
Number of votes	10	9		12

B is eliminated.
Round 3: Of the nine votes going to *B*, five now go to *A* and four go to *D*.

Candidate	*A*	*B*	*C*	*D*
Number of votes	15			16

A is eliminated, and *D* is declared the winner.
Winner: *D*. Second place: *A*. Third place: *B*. Last place: *C*.

(d) *A* versus *B*: 15 votes to 16 votes (*B* wins). *B* gets 1 point.
A versus *C*: 17 votes to 14 votes (*A* wins). *A* gets 1 point.
A versus *D*: 15 votes to 16 votes (*D* wins). *D* gets 1 point.
B versus *C*: 16 votes to 15 votes (*B* wins). *B* gets 1 point.
B versus *D*: 19 votes to 12 votes (*B* wins). *B* gets 1 point.
C versus *D*: 24 votes to 7 votes (*C* wins). *C* gets 1 point.
The final tally is 1 point for *A*, 3 points for *B*, 1 point for *C*, and 1 point for *D*.
Winner: *B*. Second place tie: *A*, *C*, and *D*.

43. (a) *A* has 40% of the first-place votes.
B has 35% of the first-place votes.
C has 25% of the first-place votes.

D has 0% of the first-place votes.
Winner *A*: Second place: *B*. Third place: *C*. Last place: *D*.

(b) We treat each 1% of voters as 1 "block" of votes.
A has $4\times40+3\times15+2\times20+1\times25=270$ points.
B has $4\times(20+15)+3\times25+2\times40+1\times0=295$ points.
C has $4\times25+3\times0+2\times0+1\times(40+20+15)=175$ points.
D has $4\times0+3\times(40+20)+2\times(25+15)+1\times0=260$ points.
Winner: *B*. Second place: *A*. Third place: *D*. Last place: *C*.

(c) Round 1:

Candidate	*A*	*B*	*C*	*D*
Number of votes	40%	35%	25%	0%

D is eliminated.
Round 2: Eliminating *D* does not affect the first-place votes of the other candidates. Using the table from Round 1, we see that candidate *C* now has the fewest number of first-place votes, so *C* is eliminated.
Round 3: The 25% of the votes going to *C* now go to *B*.

Candidate	*A*	*B*	*C*	*D*
Number of votes	40%	60%		

A is eliminated, and *B* is declared the winner.
Winner: *B*. Second place: *A*. Third place: *C*. Last place: *D*.

(d) *A* versus *B*: 40% of the votes to 60% of the votes (*B* wins). *B* gets 1 point.
A versus *C*: 75% of the votes to 25% of the votes (*A* wins). *A* gets 1 point.
A versus *D*: 55% of the votes to 45% of the votes (*A* wins). *A* gets 1 point.
B versus *C*: 75% of the votes to 25% of the votes (*B* wins). *B* gets 1 point.
B versus *D*: 60% of the votes to 40% of the votes (*B* wins). *B* gets 1 point.
C versus *D*: 25% of the votes to 75% of the votes (*D* wins). *D* gets 1 point.
The final tally is 2 points for *A*, 3 points for *B*, 0 points for *C*, and 1 point for *D*.
Winner: *B*. Second place: *A*. Third place: *D*. Last place: *C*.

45. Step 1: From the text we know that the winner using the Borda count method is *B* with 106 points.
Step 2: Removing *B* gives the following preference schedule.

Number of voters	14	10	8	4	1
1st choice	*A*	*C*	*D*	*D*	*C*
2nd choice	*C*	*D*	*C*	*C*	*D*
3rd choice	*D*	*A*	*A*	*A*	*A*

This preference schedule can be consolidated (simplified) to produce the following schedule.

Number of voters	14	11	12
1st choice	*A*	*C*	*D*
2nd choice	*C*	*D*	*C*
3rd choice	*D*	*A*	*A*

A has $3\times14+2\times0+1\times(11+12)=65$ points;
C has $3\times11+2\times(14+12)+1\times0=85$ points;
D has $3\times12+2\times11+1\times14=72$ points;
In this schedule the winner using Borda count is *C*, with 85 points. So, second place goes to *C*.

Step 3: Removing C gives the following preference schedule.

Number of voters	14	11	12
1st choice	A	D	D
2nd choice	D	A	A

A has $2\times14+1\times(11+12)=51$ points;

D has $2\times(11+12)+1\times14=60$ points;

In this schedule the winner using the Borda count method is D, with 60 points. Thus, third place goes to D, and last place goes to A.

47. Step 1: From 41(a) we know that the winner, using the plurality method, is A with 10 first-place votes.
Step 2: Removing A gives the following preference schedule.

Number of voters	10	7	9	5
1st choice	C	D	B	C
2nd choice	B	B	C	D
3rd choice	D	C	D	B

In this schedule the winner using plurality is C, with 15 first-place votes. Thus, second place goes to C.
Step 3: Removing C gives the following preference schedule.

Number of voters	19	12
1st choice	B	D
2nd choice	D	B

In this schedule the winner using plurality is B, with 19 first-place votes. Thus, third place goes to B and last place goes to D.

49. Step 1: From 41(d) we know that the winner using pairwise comparisons is B.
Step 2: Removing B gives the following preference schedule.

Number of voters	10	7	5	9
1st choice	A	D	C	C
2nd choice	C	A	A	D
3rd choice	D	C	D	A

A versus C: 17 votes to 14 votes (A wins). A gets 1 point.
A versus D: 15 votes to 16 votes (D wins). D gets 1 point.
C versus D: 24 votes to 7 votes (C wins). C gets 1 point.
The final tally is 1 point for A, 1 point for C, and 1 point for D. Since we have a three-way tie, second place goes to A, C, and D.

G. Miscellaneous

51. $\dfrac{500\times501}{2}=125,250$

53. $1+2+3+\ldots+3218+3219+3220=\dfrac{3220\times3221}{2}=5,185,810$
Combining this with the result from Exercise 51, we get
$501+502+503+\ldots+3218+3219+3220=5,185,810-125,250=5,060,560$.

55. **(a)** $1+2+3+\ldots+12+13+14=\dfrac{14\times15}{2}=105$

(b) 105 minutes = 1 hour and 45 minutes

57. Each column corresponds to a unique ordering of candidates *A*, *B*, and *C*. There are 3 choices as to which candidate is listed first, 2 choices for which candidate is listed second, and 1 choice as to which candidate is listed last. This leads to $3\times2\times1=6$ ways that the candidates can be ordered on the ballots. (These are *ABC*, *ACB*, *BAC*, *BCA*, *CAB*, *CBA*.)

59. **(a)** *A* has a majority of the first-place votes (7), so *A* is the Condorcet candidate.

(b) *A* has $4\times7+3\times2+2\times0+1\times4=38$ points;
B has $4\times4+3\times7+2\times0+1\times2=39$ points;
C has $4\times0+3\times0+2\times(7+4+2)+1\times0=26$ points;
D has $4\times2+3\times4+2\times0+1\times7=27$ points;
The winner is *B*.

(c) Removing *C* gives the following preference schedule.

Number of voters	7	4	2
1st choice	*A*	*B*	*D*
2nd choice	*B*	*D*	*A*
3rd choice	*D*	*A*	*B*

A has $3\times7+2\times2+1\times4=29$ points;
B has $3\times4+2\times7+1\times2=28$ points;
D has $3\times2+2\times4+1\times7=21$ points;
The winner is *A*.

(d) Based on (a) and (b), the Condorcet criterion and the majority criterion are violated. Based on (b) and (c), the independence of irrelevant alternatives criterion is violated. Furthermore, based on (b), the Borda count also violates the majority criterion since *A* has a majority of the first-place votes but does not win the election.

JOGGING

61. Suppose the two candidates are *A* and *B* and that *A* gets *a* first-place votes and *B* gets *b* first-place votes and suppose that $a > b$. Then *A* has a majority of the votes and the preference schedule is

Number of voters	*a*	*b*
1st choice	*A*	*B*
2nd choice	*B*	*A*

It is clear that candidate *A* wins the election under the plurality method, the plurality-with-elimination method, and the method of pairwise comparisons. Under the Borda count method, *A* gets $2a + b$ points while *B* gets $2b + a$ points. Since $a > b$, $2a + b > 2b + a$ and so again *A* wins the election.

63. If *X* is the winner of an election using the plurality method and, in a reelection, the only changes in the ballots are changes that only favor *X*, then no candidate other than *X* can increase his/her first-place votes and so *X* is still the winner of the election.

65. If *X* is the winner of an election using the method of pairwise comparisons and, in a reelection, the only changes in the ballots are changes that favor *X* and only favor *X*, then candidate *X* will still win every pairwise comparison that he/she won in the original election and possibly even some new ones — while no other candidate will win any new pairwise comparisons (since there were no changes favorable to any other candidate). That is, if *X* has *A* wins in the original election and A' wins in the reelection, then $A' \geq A$. Similarly, if *Y* has *B* wins in the original election and B' wins in the reelection, then $B' \leq B$. So, we have $A' - A \geq 0 \geq B' - B$ and *X* will gain more points than any other candidate will gain and hence will remain the winner of the election.

67. **(a)** Suppose a candidate, *C*, gets v_1 first-place votes, v_2 second-place votes, v_3 third-place votes, …, v_N *N*th-place votes. For this candidate $p = v_1 N + v_2(N-1) + v_3(N-2) + \ldots + v_{N-1}\cdot 2 + v_N \cdot 1$, and $r = v_1 \cdot 1 + v_2 \cdot 2 + v_3 \cdot 3 + \ldots + v_{N-1}(N-1) + v_N N$.

So, $p + r = v_1(N+1) + v_2(N+1) + v_3(N+1) + \ldots + v_{N-1}(N+1) + v_N(N+1)$

$= (v_1 + v_2 + \ldots + v_N)(N+1)$

$= k(N+1)$

(b) Suppose candidates C_1 and C_2 receive p_1 and p_2 points respectively using the Borda count as originally described in the chapter and r_1 and r_2 points under the variation described in this exercise. Then if $p_1 < p_2$, we have $-p_1 > -p_2$ and so $k(N+1) - p_1 > k(N+1) - p_2$ which implies [using part (a)], $r_1 > r_2$. Consequently the relative ranking of the candidates is not changed.

69. **(a)** Each of the voters gave Ohio State 25 points, so we compute 1625/25 to find 65 voters in the poll.

(b) Florida: Let x = the number of second-place votes. Then $1529 = 24 \cdot x + 23 \cdot (65 - x)$ and so $x = 34$. This leaves 65 – 34 = 31 third-place votes.

(c) Michigan: Let y = the number of second-place votes. Then $1526 = 24 \times y + 23 \times (65 - y)$ and so $y = 31$. This leaves 65 – 31 = 34 third-place votes.

71. By looking at Dwayne Wade's vote totals, it is clear that 1 point is awarded for each third-place vote. Let x = points awarded for each second-place vote. Then, $3x + 108 = 117$ gives $x = 3$ points for each second-place vote. Next, let y = points awarded for each first-place vote. Looking at LeBron James' votes, we see that $78y + 39 \times 3 + 1 \times 1 = 508$. Solving this equation gives $y = 5$ points for each first-place vote.

73. **(a)** C ; A is eliminated first, D is eliminated next, and then C beats B.

(b) A is a Condorcet candidate but is eliminated in the first round.

Number of voters	10	6	6	3	3
1st choice	*B*	*A*	*A*	*D*	*C*
2nd choice	*C*	*B*	*C*	*A*	*A*
3rd choice	*D*	*D*	*B*	*C*	*B*
4th choice	*A*	*C*	*D*	*B*	*D*

(c) B wins under the Coombs method. However, if 8 voters move B from their 3rd choice to their 2nd choice, then C wins.

Number of voters	10	8	7	4
1st choice	*B*	*C*	*C*	*A*
2nd choice	*A*	*A*	*B*	*B*
3rd choice	*C*	*B*	*A*	*C*

Chapter 2

WALKING

A. Weighted Voting Systems

1. (a) There are 6 players.

(b) 30 + 28 + 22 + 15 + 7 + 6 = 108 votes.

(c) 28; this is the second number to the right of the colon (:).

(d) $\frac{65}{108} \approx 0.60185$; So, rounded to the next whole percent, 61% of the votes are needed to pass a motion.

3. (a) The quota must be more than half of the total number of votes. This system has 10 + 6 + 5 + 4 + 2 = 27 total votes. $\frac{1}{2} \times 27 = 13.5$, so the smallest value q can take is 14.

(b) The largest value q can take is 27, the total number of votes.

(c) $\frac{2}{3} \times 27 = 18$, so the value of the quota q would be 18.

(d) The value of the quota q would be strictly larger than 18. That is, 19.

5. To determine the number of votes each player has, write the weighted voting system as [49: 4x, 2x, x, x].

(a) If the quota is defined as a simple majority of the votes, then x is the largest integer satisfying $49 > \frac{4x+2x+x+x}{2}$ which means that $8x < 98$ or $x < 12.25$. So, $x = 12$ and the system can be described as [49: 48, 24, 12, 12].

(b) If the quota is defined as more than two-thirds of the votes, then x is the largest integer satisfying $49 > \frac{2(4x+2x+x+x)}{3}$ which means that $16x < 147$ or $x < 9.1875$. So, $x = 9$ and the system can be described as [49: 36, 18, 9, 9].

(c) If the quota is defined as more than three-fourths of the votes, then x is the largest integer satisfying $49 > \frac{3(4x+2x+x+x)}{4}$ which means that $24x < 196$ or $x < 8.167$. So, $x = 8$ and the system can be described as [49: 32, 16, 8, 8].

7. (a) P_1, with more votes than the quota, is a dictator; the other players are dummies.

(b) P_1 has veto power since the other players cannot pass a motion without P_1. P_4 is a dummy since the quota is even and all players other than P_4 (having one vote) have an even number of votes.

(c) P_1 and P_2 have veto power since the other players cannot pass a motion without P_2 (and hence certainly without the votes of player P_1 either). P_3 and P_4 are dummies since the only way a motion passes is if P_1 and P_2 vote in favor of it.

9. (a) All three players will have veto power exactly when it takes all three players to pass a motion. The smallest value for such a quota is when it is one more than (strictly greater than) the number of votes that all but the weakest player (P_3) control. In this case, all players have veto power when $q > 8 + 4 = 12$. The smallest such value is, of course, $q = 13$.

(b) P_2 has veto power when the quota q is greater than the number of votes the other players control. In this case, since P_1 and P_3 control 8 + 2 = 10 votes, $q > 10$ must hold for P_2 to have veto power. For P_3 to not have veto power, the other players must be able to pass a motion without the votes of P_3. That is, the quota q must satisfy $q \leq 12$. The smallest such value is $q = 11$.

(c) In order for P_3 to be a dummy, $q > 10$ must hold (otherwise P_3 is a critical player in a two-player coalition). Since P_3 is a dummy when $q = 11$, that is the smallest such value.

B. Banzhaf Power

11. (a) $6 + 4 = 10$

(b) $\{P_1, P_2\}$, $\{P_1, P_3\}$, $\{P_1, P_2, P_3\}$, $\{P_1, P_2, P_4\}$, $\{P_1, P_3, P_4\}$, $\{P_2, P_3, P_4\}$, $\{P_1, P_2, P_3, P_4\}$

(c) P_1 only

(d)

Winning Coalitions	Critical players
$\{P_1, P_2\}$	P_1, P_2
$\{P_1, P_3\}$	P_1, P_3
$\{P_1, P_2, P_3\}$	P_1
$\{P_1, P_2, P_4\}$	P_1, P_2
$\{P_1, P_3, P_4\}$	P_1, P_3
$\{P_2, P_3, P_4\}$	P_2, P_3, P_4
$\{P_1, P_2, P_3, P_4\}$	none

P_1 is critical five times; P_2 is critical three times; P_3 is critical three times; P_4 is critical one time. The total number of times the players are critical is 5 + 3 + 3 + 1 = 12. The Banzhaf power distribution is $\beta_1 = \frac{5}{12} = 41\frac{2}{3}\%$; $\beta_2 = \frac{3}{12} = 25\%$; $\beta_3 = \frac{3}{12} = 25\%$; $\beta_4 = \frac{1}{12} = 8\frac{1}{3}\%$.

13. (a)

Winning Coalitions	Critical players
$\{P_1, P_2\}$	P_1, P_2
$\{P_1, P_3\}$	P_1, P_3
$\{P_1, P_2, P_3\}$	P_1

P_1 is critical three times; P_2 is critical one time; P_3 is critical one time. The total number of times the players are critical is 3 + 1 + 1 = 5. The Banzhaf power distribution is $\beta_1 = \frac{3}{5} = 60\%$; $\beta_2 = \frac{1}{5} = 20\%$; $\beta_3 = \frac{1}{5} = 20\%$.

(b)

Winning Coalitions	Critical players
$\{P_1, P_2\}$	P_1, P_2
$\{P_1, P_3\}$	P_1, P_3
$\{P_1, P_2, P_3\}$	P_1

P_1 is critical three times; P_2 is critical one time; P_3 is critical one time. The total number of times the players are critical is 3 + 1 + 1 = 5. The Banzhaf power distribution is $\beta_1 = \frac{3}{5} = 60\%$; $\beta_2 = \frac{1}{5} = 20\%$; $\beta_3 = \frac{1}{5} = 20\%$. The answers to (a) and (b) are the same.

15. (a)

Winning Coalitions	Critical players
$\{P_1, P_2, P_3\}$	P_1, P_2, P_3
$\{P_1, P_2, P_4\}$	P_1, P_2, P_4
$\{P_1, P_2, P_5\}$	P_1, P_2, P_5
$\{P_1, P_3, P_4\}$	P_1, P_3, P_4
$\{P_1, P_2, P_3, P_4\}$	P_1
$\{P_1, P_2, P_3, P_5\}$	P_1, P_2
$\{P_1, P_2, P_4, P_5\}$	P_1, P_2
$\{P_1, P_3, P_4, P_5\}$	P_1, P_3, P_4
$\{P_2, P_3, P_4, P_5\}$	P_2, P_3, P_4, P_5
$\{P_1, P_2, P_3, P_4, P_5\}$	none

P_1 is critical eight times; P_2 is critical six times; P_3 is critical four times; P_4 is critical four times; P_5 is critical two times. The total number of times the players are critical is 8 + 6 + 4 + 4 + 2 = 24. The Banzhaf power distribution is $\beta_1 = \frac{8}{24} = 33\frac{1}{3}\%$; $\beta_2 = \frac{6}{24} = 25\%$; $\beta_3 = \frac{4}{24} = 16\frac{2}{3}\%$; $\beta_4 = \frac{4}{24} = 16\frac{2}{3}\%$; $\beta_5 = \frac{2}{24} = 8\frac{1}{3}\%$.

(b) The quota is one more than in (a), so some winning coalitions may now be losing coalitions. For the ones that are still winning, any players that were critical in (a) will still be critical, and there may be additional critical players. A quick check shows that $\{P_1, P_2, P_5\}$, $\{P_1, P_3, P_4\}$, and $\{P_2, P_3, P_4, P_5\}$ are now losing coalitions (they all have exactly 10 votes).

Winning Coalitions	Critical players
$\{P_1, P_2, P_3\}$	P_1, P_2, P_3
$\{P_1, P_2, P_4\}$	P_1, P_2, P_4
$\{P_1, P_2, P_3, P_4\}$	P_1, P_2
$\{P_1, P_2, P_3, P_5\}$	P_1, P_2, P_3
$\{P_1, P_2, P_4, P_5\}$	P_1, P_2, P_4
$\{P_1, P_3, P_4, P_5\}$	P_1, P_3, P_4, P_5
$\{P_1, P_2, P_3, P_4, P_5\}$	P_1

P_1 is critical seven times; P_2 is critical five times; P_3 is critical three times; P_4 is critical three times; P_5 is critical one time. The total number of times the players are critical is 7 + 5 + 3 + 3 + 1 = 19. The Banzhaf power distribution is $\beta_1 = \frac{7}{19} = 36\frac{13}{19}\%$; $\beta_2 = \frac{5}{19} = 26\frac{6}{19}\%$; $\beta_3 = \frac{3}{19} = 15\frac{15}{19}\%$; $\beta_4 = \frac{3}{19} = 15\frac{15}{19}\%$; $\beta_5 = \frac{1}{19} = 5\frac{5}{19}\%$.

17. (a) P_1 is a dictator with 8 votes (the quota). So, the Banzhaf power distribution is $\beta_1 = 1$; $\beta_2 = 0$; $\beta_3 = 0$; $\beta_4 = 0$.

(b)

Winning Coalitions	Critical players
$\{P_1, P_2\}$	P_1, P_2
$\{P_1, P_3\}$	P_1, P_3
$\{P_1, P_4\}$	P_1, P_4
$\{P_1, P_2, P_3\}$	P_1
$\{P_1, P_2, P_4\}$	P_1
$\{P_1, P_3, P_4\}$	P_1
$\{P_1, P_2, P_3, P_4\}$	P_1

P_1 is critical seven times; P_2 is critical one time; P_3 is critical one time; P_4 is critical one time. The total number of times all players are critical is 7 + 1 + 1 + 1 = 10. The Banzhaf power distribution is $\beta_1 = \frac{7}{10} = 70\%$; $\beta_2 = \frac{1}{10} = 10\%$; $\beta_3 = \frac{1}{10} = 10\%$; $\beta_4 = \frac{1}{10} = 10\%$.

(c) This situation is like (b) with the following exceptions: $\{P_1, P_4\}$ is now a losing coalition; P_1 and P_2 are both critical in $\{P_1, P_2, P_4\}$; P_1 and P_3 are both critical in $\{P_1, P_3, P_4\}$. Now P_1 is critical six times; P_2 is critical two times; P_3 is critical two times; P_4 is never critical. The total number of times all players are critical is 6 + 2 + 2 = 10. The Banzhaf power distribution is $\beta_1 = \frac{6}{10} = 60\%$; $\beta_2 = \frac{2}{10} = 20\%$; $\beta_3 = \frac{2}{10} = 20\%$; $\beta_4 = 0$.

(d)

Winning Coalitions	Critical players
$\{P_1, P_2\}$	P_1, P_2
$\{P_1, P_2, P_3\}$	P_1, P_2
$\{P_1, P_2, P_4\}$	P_1, P_2
$\{P_1, P_2, P_3, P_4\}$	P_1, P_2

P_1 and P_2 are each critical in every winning coalition, and P_3 and P_4 are never critical. The Banzhaf power distribution is $\beta_1 = \frac{1}{2} = 50\%$; $\beta_2 = \frac{1}{2} = 50\%$; $\beta_3 = 0$; $\beta_4 = 0$.

(e)

Winning Coalitions	Critical players
$\{P_1, P_2, P_3\}$	P_1, P_2, P_3
$\{P_1, P_2, P_3, P_4\}$	P_1, P_2, P_3

P_1, P_2, and P_3 are each critical in every winning coalition. P_4 is never critical. The Banzhaf power distribution is $\beta_1 = \frac{1}{3} = 33\frac{1}{3}\%$; $\beta_2 = \frac{1}{3} = 33\frac{1}{3}\%$; $\beta_3 = \frac{1}{3} = 33\frac{1}{3}\%$; $\beta_4 = 0$.

19. (a) A player is critical in a coalition if that coalition without the player is not on the list of winning coalitions. So, in this case, the critical players are underlined below.

$\{\underline{P_1}, \underline{P_2}\}$, $\{\underline{P_1}, \underline{P_3}\}$, $\{\underline{P_1}, P_2, P_3\}$

(b) P_1 is critical three times; P_2 is critical one time; P_3 is critical one time. The total number of times all players are critical is 3 + 1 + 1 = 5. The Banzhaf power distribution is $\beta_1 = \frac{3}{5} = 60\%$; $\beta_2 = \frac{1}{5} = 20\%$; $\beta_3 = \frac{1}{5} = 20\%$.

21. (a) $\{\underline{P_1}, \underline{P_2}\}$, $\{\underline{P_1}, \underline{P_3}\}$, $\{\underline{P_2}, \underline{P_3}\}$, $\{P_1, P_2, P_3\}$, $\{\underline{P_1}, \underline{P_2}, P_4\}$, $\{\underline{P_1}, \underline{P_2}, P_5\}$, $\{\underline{P_1}, \underline{P_2}, P_6\}$, $\{\underline{P_1}, \underline{P_3}, P_4\}$, $\{\underline{P_1}, \underline{P_3}, P_5\}$, $\{\underline{P_1}, \underline{P_3}, P_6\}$, $\{\underline{P_2}, \underline{P_3}, P_4\}$, $\{\underline{P_2}, \underline{P_3}, P_5\}$, $\{\underline{P_2}, \underline{P_3}, P_6\}$

(b) $\{\underline{P_1}, \underline{P_2}, P_4\}$, $\{\underline{P_1}, \underline{P_3}, P_4\}$, $\{\underline{P_2}, \underline{P_3}, P_4\}$, $\{P_1, P_2, P_3, P_4\}$, $\{\underline{P_1}, \underline{P_2}, P_4, P_5\}$, $\{\underline{P_1}, \underline{P_2}, P_4, P_6\}$, $\{\underline{P_1}, \underline{P_3}, P_4, P_5\}$, $\{\underline{P_1}, \underline{P_3}, P_4, P_6\}$, $\{\underline{P_2}, \underline{P_3}, P_4, P_5\}$, $\{\underline{P_2}, \underline{P_3}, P_4, P_6\}$, $\{P_1, P_2, P_3, P_4, P_5\}$, $\{P_1, P_2, P_3, P_4, P_6\}$, $\{\underline{P_1}, \underline{P_2}, P_4, P_5, P_6\}$, $\{\underline{P_1}, \underline{P_3}, P_4, P_5, P_6\}$, $\{\underline{P_2}, \underline{P_3}, P_4, P_5, P_6\}$, $\{P_1, P_2, P_3, P_4, P_5, P_6\}$

(c) P_4 is never a critical player since every time it is part of a winning coalition, that coalition is a winning coalition without P_4 as well. So, $\beta_4 = 0$.

(d) A similar argument to that used in part (c) shows that P_5 and P_6 are also dummies. One could also argue that any player with fewer votes than P_4, a dummy, will also be a dummy. So, P_4, P_5, and P_6 will never be critical -- they all have zero power.

The only winning coalitions with only two players are $\{P_1, P_2\}$, $\{P_1, P_3\}$, and $\{P_2, P_3\}$; and both players are critical in each of those coalitions. All other winning coalitions consist of one of these coalitions plus additional players, and the only critical players will be the ones from the two-player coalition. So P_1, P_2, and P_3 will be critical in every winning coalition they are in, and they will all be in the same number of winning coalitions, so they all have the same power. Thus, the Banzhaf power distribution is

$$\beta_1 = \frac{1}{3} = 33\frac{1}{3}\%;\ \beta_2 = \frac{1}{3} = 33\frac{1}{3}\%;\ \beta_3 = \frac{1}{3} = 33\frac{1}{3}\%;\ \beta_4 = 0;\ \beta_5 = 0;\ \beta_6 = 0.$$

23. (a) $\{\underline{A}, \underline{B}\}, \{\underline{A}, \underline{C}\}, \{\underline{B}, \underline{C}\}, \{A, B, C\}, \{\underline{A}, \underline{B}, D\}, \{\underline{A}, \underline{C}, D\}, \{\underline{B}, \underline{C}, D\}, \{A, B, C, D\}$

(b) *A*, *B*, and *C* have Banzhaf power index of 1/3 each; *D* is a dummy. (*D* is never a critical player, and the other three clearly have equal power.)

C. Shapley-Shubik Power

25. (a) There are 3! = 6 sequential coalitions of the three players.
$\langle P_1, \underline{P_2}, P_3\rangle, \langle P_1, \underline{P_3}, P_2\rangle, \langle P_2, \underline{P_1}, P_3\rangle, \langle P_2, P_3, \underline{P_1}\rangle, \langle P_3, \underline{P_1}, P_2\rangle, \langle P_3, P_2, \underline{P_1}\rangle$

(b) P_1 is pivotal four times; P_2 is pivotal one time; P_3 is pivotal one time. The Shapley-Shubik power distribution is $\sigma_1 = \frac{4}{6} = 66\frac{2}{3}\%;\ \sigma_2 = \frac{1}{6} = 16\frac{2}{3}\%;\ \sigma_3 = \frac{1}{6} = 16\frac{2}{3}\%.$

27. (a) Since P_1 is a dictator, $\sigma_1 = 1$, $\sigma_2 = 0$, $\sigma_3 = 0$, and $\sigma_4 = 0$.

(b) There are 4! = 24 sequential coalitions of the four players. The pivotal player in each coalition is underlined.
$\langle P_1, \underline{P_2}, P_3, P_4\rangle, \langle P_1, \underline{P_2}, P_4, P_3\rangle, \langle P_1, \underline{P_3}, P_2, P_4\rangle, \langle P_1, \underline{P_3}, P_4, P_2\rangle,$
$\langle P_1, P_4, \underline{P_2}, P_3\rangle, \langle P_1, P_4, \underline{P_3}, P_2\rangle, \langle P_2, \underline{P_1}, P_3, P_4\rangle, \langle P_2, \underline{P_1}, P_4, P_3\rangle,$
$\langle P_2, P_3, \underline{P_1}, P_4\rangle, \langle P_2, P_3, P_4, \underline{P_1}\rangle, \langle P_2, P_4, \underline{P_1}, P_3\rangle, \langle P_2, P_4, P_3, \underline{P_1}\rangle,$
$\langle P_3, \underline{P_1}, P_2, P_4\rangle, \langle P_3, \underline{P_1}, P_4, P_2\rangle, \langle P_3, P_2, \underline{P_1}, P_4\rangle, \langle P_3, P_2, P_4, \underline{P_1}\rangle,$
$\langle P_3, P_4, \underline{P_1}, P_2\rangle, \langle P_3, P_4, P_2, \underline{P_1}\rangle, \langle P_4, P_1, \underline{P_2}, P_3\rangle, \langle P_4, P_1, \underline{P_3}, P_2\rangle,$
$\langle P_4, P_2, \underline{P_1}, P_3\rangle, \langle P_4, P_2, P_3, \underline{P_1}\rangle, \langle P_4, P_3, \underline{P_1}, P_2\rangle, \langle P_4, P_3, P_2, \underline{P_1}\rangle$

P_1 is pivotal 16 times; P_2 is pivotal 4 times; P_3 is pivotal 4 times; P_4 is pivotal 0 times.

The Shapley-Shubik power distribution is $\sigma_1 = \frac{16}{24} = \frac{4}{6}$; $\sigma_2 = \frac{4}{24} = \frac{1}{6}$; $\sigma_3 = \frac{4}{24} = \frac{1}{6}$; $\sigma_4 = 0$.

(c) The only way a motion will pass is if P_1 and P_2 both support it. In fact, the second of these players that appears in a sequential coalition will be the pivotal player in that coalition. It follows that $\sigma_1 = \frac{12}{24} = \frac{1}{2}$, $\sigma_2 = \frac{12}{24} = \frac{1}{2}$, $\sigma_3 = 0$, and $\sigma_4 = 0$.

(d) Because the quota is so high, the only way a motion will pass is if P_1, P_2, and P_3 all support it. In fact, the third of these players that appears in a sequential coalition will always be the pivotal player in that coalition. It follows that $\sigma_1 = \frac{8}{24} = \frac{1}{3}$, $\sigma_2 = \frac{8}{24} = \frac{1}{3}$, $\sigma_3 = \frac{8}{24} = \frac{1}{3}$, and $\sigma_4 = 0$.

29. (a) There are 4! = 24 sequential coalitions of the four players. In each coalition the pivotal player is underlined.

$< P_1, \underline{P_2}, P_3, P_4 >, < P_1, \underline{P_2}, P_4, P_3 >, < P_1, \underline{P_3}, P_2, P_4 >, < P_1, \underline{P_3}, P_4, P_2 >,$

$< P_1, P_4, \underline{P_2}, P_3 >, < P_1, P_4, \underline{P_3}, P_2 >, < P_2, \underline{P_1}, P_3, P_4 >, < P_2, \underline{P_1}, P_4, P_3 >,$

$< P_2, P_3, \underline{P_1}, P_4 >, < P_2, P_3, \underline{P_4}, P_1 >, < P_2, P_4, \underline{P_1}, P_3 >, < P_2, P_4, \underline{P_3}, P_1 >,$

$< P_3, \underline{P_1}, P_2, P_4 >, < P_3, \underline{P_1}, P_4, P_2 >, < P_3, P_2, \underline{P_1}, P_4 >, < P_3, P_2, \underline{P_4}, P_1 >,$

$< P_3, P_4, \underline{P_1}, P_2 >, < P_3, P_4, \underline{P_2}, P_1 >, < P_4, P_1, \underline{P_2}, P_3 >, < P_4, P_1, \underline{P_3}, P_2 >,$

$< P_4, P_2, \underline{P_1}, P_3 >, < P_4, P_2, \underline{P_3}, P_1 >, < P_4, P_3, \underline{P_1}, P_2 >, < P_4, P_3, \underline{P_2}, P_1 >$

P_1 is pivotal in 10 coalitions; P_2 is pivotal in six coalitions; P_3 is pivotal in six coalitions; P_4 is pivotal in two coalitions. The Shapley-Shubik power distribution is $\sigma_1 = \frac{10}{24} = 41\frac{2}{3}\%$;

$\sigma_2 = \frac{6}{24} = 25\%$; $\sigma_3 = \frac{6}{24} = 25\%$; $\sigma_4 = \frac{2}{24} = 8\frac{1}{3}\%$.

(b) This is the same situation as in (a) – there is essentially no difference between 51 and 59 because the players' votes are all multiples of 10. The Shapley-Shubik power distribution is thus still $\sigma_1 = \frac{10}{24} = 41\frac{2}{3}\%$; $\sigma_2 = \frac{6}{24} = 25\%$; $\sigma_3 = \frac{6}{24} = 25\%$; $\sigma_4 = \frac{2}{24} = 8\frac{1}{3}\%$.

(c) This is also the same situation as in (a) – any time a group of players has 51 votes, they must have 60 votes. The Shapley-Shubik power distribution is thus still $\sigma_1 = \frac{10}{24} = 41\frac{2}{3}\%$; $\sigma_2 = \frac{6}{24} = 25\%$;

$\sigma_3 = \frac{6}{24} = 25\%$; $\sigma_4 = \frac{2}{24} = 8\frac{1}{3}\%$.

31. (a) There are 3! = 6 sequential coalitions of the three players.

$< P_1, \underline{P_2}, P_3 >, < P_1, \underline{P_3}, P_2 >, < P_2, \underline{P_1}, P_3 >, < P_2, P_3, \underline{P_1} >, < P_3, \underline{P_1}, P_2 >, < P_3, P_2, \underline{P_1} >$

(The second player listed will always be pivotal unless the first two players listed are P_2 and P_3 since that is not a winning coalition. In that case, P_1 is pivotal.)

(b) P_1 is pivotal four times; P_2 is pivotal one time; P_3 is pivotal one time. The Shapley-Shubik power distribution is $\sigma_1 = \frac{4}{6} = 66\frac{2}{3}\%$; $\sigma_2 = \frac{1}{6} = 16\frac{2}{3}\%$; $\sigma_3 = \frac{1}{6} = 16\frac{2}{3}\%$.

33. *D* will never be pivotal and has no power. Each of the other players will be pivotal in the same number of coalitions, so they will all have equal power. Thus, the Shapley-Shubik power distribution is $A: \frac{1}{3} = 33\frac{1}{3}\%$; $B: \frac{1}{3} = 33\frac{1}{3}\%$; $C: \frac{1}{3} = 33\frac{1}{3}\%$; $D: 0$.

D. Miscellaneous

35. (a) $13! = 6{,}227{,}020{,}800$

(b) $18! = 6{,}402{,}373{,}705{,}728{,}000 \approx 6.402374 \times 10^{15}$

(c) $25! = 15{,}511{,}210{,}043{,}330{,}985{,}984{,}000{,}000 \approx 1.551121 \times 10^{25}$

(d) There are 25! sequential coalitions of 25 players.

$$25! \text{ sequential coaltions} \times \frac{1 \text{ second}}{1{,}000{,}000{,}000{,}000 \text{ sequential coalitions}} \times \frac{1 \text{ hour}}{3600 \text{ seconds}} \times \frac{1 \text{ day}}{24 \text{ hours}} \times \frac{1 \text{ year}}{365 \text{ days}}$$

$\approx 491{,}857$ years

37. (a) $10! = 10 \times 9 \times 8 \times \ldots \times 3 \times 2 \times 1$

$= 10 \times 9!$

So, $9! = \frac{10!}{10} = \frac{3{,}628{,}800}{10} = 362{,}880$.

(b) $11! = 11 \times 10 \times 9 \times \ldots \times 3 \times 2 \times 1 = 11 \times 10!$

So, $\frac{11!}{10!} = \frac{11 \times 10!}{10!} = 11$.

(c) $11! = 11 \times 10 \times (9 \times \ldots \times 3 \times 2 \times 1) = 11 \times 10 \times 9!$

So, $\frac{11!}{9!} = \frac{11 \times 10 \times 9!}{9!} = 11 \times 10 = 110$.

(d) $\frac{9!}{6!} = \frac{9 \times 8 \times 7 \times 6!}{6!} = 9 \times 8 \times 7 = 504$

(e) $\frac{101!}{99!} = \frac{101 \times 100 \times 99!}{99!} = 101 \times 100 = 10{,}100$

39. (a) $\frac{9! + 11!}{10!} = \frac{9!}{10!} + \frac{11!}{10!} = \frac{9!}{10 \times 9!} + \frac{11 \times 10!}{10!} = \frac{1}{10} + 11 = 11.1$

(b) $\frac{101! + 99!}{100!} = \frac{101!}{100!} + \frac{99!}{100!} = \frac{101 \times 100!}{100!} + \frac{99!}{100 \times 99!} = 101 + \frac{1}{100} = 101.01$

41. (a) $2^6 - 1 = 63$ coalitions

(b) There are $2^5 - 1 = 31$ coalitions of the remaining five players P_2, P_3, P_4, P_5, and P_6. These are exactly those coalitions that do not include P_1.

(c) As in (b), there are $2^5 - 1 = 31$ coalitions of the remaining five players P_1, P_2, P_4, P_5, and P_6. These are exactly those coalitions that do not include P_3.

(d) There are $2^4 - 1 = 15$ coalitions of the remaining four players P_2, P_4, P_5, P_6. These are exactly those coalitions that do not include P_1 or P_3.

(e) 63 – 15 = 48 coalitions include P_1 and P_3.

43. (a) $5! = 120$ sequential coalitions

(b) There are $4! = 24$ sequential coalitions of the remaining four players P_2, P_3, P_4 and P_5. These are exactly those coalitions that have P_1 as the first player.

(c) As in (b), there are $4! = 24$ sequential coalitions of the remaining four players P_2, P_3, P_4 and P_5. These are exactly those coalitions that have P_1 as the last player.

(d) 120 – 24 = 96 sequential coalitions do not have P_1 as the first player.

45. (a) First, note that P_1 is pivotal in all sequential coalitions except when it is the first player. By 43(d), P_1 is pivotal in 96 of the 120 sequential coalitions.

(b) Based on the results of (a), $\sigma_1 = \frac{96}{120} = \frac{4}{5}$.

(c) Since players P_2, P_3, P_4 and P_5 share the remaining power equally, $\sigma_1 = \frac{96}{120} = \frac{4}{5}$,

$\sigma_2 = \sigma_3 = \sigma_4 = \sigma_5 = \frac{6}{120} = \frac{1}{20}$.

JOGGING

47. Write the weighted voting system as [*q*: 8*x*, 4*x*, 2*x*, *x*]. The total number of votes is 15*x*. If *x* is even, then so is 15*x*. Since the quota is a simple majority, $q = \frac{15x}{2} + 1$. But, $\frac{15x}{2} + 1 = 7.5x + 1 \le 8x$ since $x \ge 2$. So, P_1 (having 8*x* votes) is a dictator. If *x* is odd, then so is 15*x*. Since the quota is a simple majority, $q = \frac{15x+1}{2}$. But, $\frac{15x+1}{2} = 7.5x + \frac{1}{2} \le 8x$ since $x \ge 1$. So, again, P_1 (having 8*x* votes) is a dictator.

49. Since P_2 is a pivotal player in the sequential coalition $< P_1, P_2, P_3 >$, it is clear that $\{P_1, P_2\}$ and $\{P_1, P_2, P_3\}$ are winning coalitions, but $\{P_1\}$ is a losing coalition. Similarly, the location of each pivotal player in each sequential coalition gives information as shown in the chart below.

Pivotal Player and Sequential Coalition	Winning Coalitions	Losing Coalitions
$< P_1, \underline{P_2}, P_3 >$	$\{P_1, P_2\}$, $\{P_1, P_2, P_3\}$	$\{P_1\}$
$< P_1, \underline{P_3}, P_2 >$	$\{P_1, P_3\}$, $\{P_1, P_2, P_3\}$	$\{P_1\}$
$< P_2, \underline{P_1}, P_3 >$	$\{P_1, P_2\}$, $\{P_1, P_2, P_3\}$	$\{P_2\}$
$< P_2, P_3, \underline{P_1} >$	$\{P_1, P_2, P_3\}$	$\{P_2\}$, $\{P_3\}$, $\{P_2, P_3\}$
$< P_3, \underline{P_1}, P_2 >$	$\{P_1, P_3\}$, $\{P_1, P_2, P_3\}$	$\{P_3\}$
$< P_3, P_2, \underline{P_1} >$	$\{P_1, P_2, P_3\}$	$\{P_2\}$, $\{P_3\}$, $\{P_2, P_3\}$

Based on this, the winning coalitions with critical players underlined are $\{\underline{P_1}, \underline{P_2}\}$, $\{\underline{P_1}, \underline{P_3}\}$, and $\{\underline{P_1}, P_2, P_3\}$.

It follows that the Banzhaf power indices are $\beta_1 = \frac{3}{5} = 60\%$; $\beta_2 = \frac{1}{5} = 20\%$; $\beta_3 = \frac{1}{5} = 20\%$.

51. **(a)** [4: 2, 1, 1, 1] or [9: 5, 2, 2, 2] are among the possible answers.

(b) The sequential coalitions (with pivotal players underlined) are:
$< H, A_1, \underline{A_2}, A_3 >$, $< H, A_1, \underline{A_3}, A_2 >$, $< H, A_2, \underline{A_1}, A_3 >$, $< H, A_2, \underline{A_3}, A_1 >$,
$< H, A_3, \underline{A_1}, A_2 >$, $< H, A_3, \underline{A_2}, A_1 >$, $< A_1, H, \underline{A_2}, A_3 >$, $< A_1, H, \underline{A_3}, A_2 >$,
$< A_1, A_2, \underline{H}, A_3 >$, $< A_1, A_2, A_3, \underline{H} >$, $< A_1, A_3, \underline{H}, A_2 >$, $< A_1, A_3, A_2, \underline{H} >$,
$< A_2, H, \underline{A_1}, A_3 >$, $< A_2, H, \underline{A_3}, A_1 >$, $< A_2, A_1, \underline{H}, A_3 >$, $< A_2, A_1, A_3, \underline{H} >$,
$< A_2, A_3, \underline{H}, A_1 >$, $< A_2, A_3, A_1, \underline{H} >$, $< A_3, H, \underline{A_1}, A_2 >$, $< A_3, H, \underline{A_2}, A_1 >$,
$< A_3, A_1, \underline{H}, A_2 >$, $< A_3, A_1, A_2, \underline{H} >$, $< A_3, A_2, \underline{H}, A_1 >$, $< A_3, A_2, A_1, \underline{H} >$.
H is pivotal in 12 coalitions; A_1 is pivotal in four coalitions; A_2 is pivotal in four coalitions; A_3 is pivotal in four coalitions. The Shapley-Shubik power distribution is $\sigma_H = \frac{12}{24} = \frac{1}{2}$;
$\sigma_{A_1} = \sigma_{A_2} = \sigma_{A_3} = \frac{4}{24} = \frac{1}{6}$.

53. **(a)** Suppose that a winning coalition that contains *P* is not a winning coalition without *P*. Then *P* would be a critical player in that coalition, contradicting the fact that *P* is a dummy.

(b) *P* is a dummy $\Leftrightarrow$ *P* is never critical $\Leftrightarrow$ the numerator of its Banzhaf power index is 0 $\Leftrightarrow$ it's Banzhaf power index is 0.

(c) Suppose *P* is not a dummy. Then, *P* is critical in some winning coalition. Let *S* denote the other players in that winning coalition. The sequential coalition with the players in *S* first (in any order), followed by *P* and then followed by the remaining players has *P* as its pivotal player. Thus, *P*'s Shapley-Shubik power index is not 0. Conversely, if *P*'s Shapley-Shubik power index is not 0, then *P* is pivotal in some sequential coalition. A coalition consisting of *P* together with the players preceding *P* in that sequential coalition is a winning coalition and *P* is a critical player in it. Thus, *P* is not a dummy.

55. **(a)** The quota must be at least half of the total number of votes and not more than the total number of votes. $7 \le q \le 13$.

(b) For $q = 7$ or $q = 8$, P_1 is a dictator because $\{P_1\}$ is a winning coalition.

(c) For $q = 9$, only P_1 has veto power since P_2 and P_3 together have just 5 votes.

(d) For $10 \le q \le 12$, both P_1 and P_2 have veto power since no motion can pass without both of their votes. For $q = 13$, all three players have veto power.

(e) For $q = 7$ or $q = 8$, both P_2 and P_3 are dummies because P_1 is a dictator. For $10 \le q \le 12$, P_3 is a dummy since all winning coalitions contain $\{P_1, P_2\}$ which is itself a winning coalition.

57. **(a)** [24: 14, 8, 6, 4] is just [12: 7, 4, 3, 2] with each value multiplied by 2. Both have Banzhaf power distribution $P_1 : \frac{2}{5} = 40\%$; $P_2 : \frac{1}{5} = 20\%$; $P_3 : \frac{1}{5} = 20\%$; $P_4 : \frac{1}{5} = 20\%$.

(b) In the weighted voting system $[q: w_1, w_2, \ldots, w_N]$, if P_k is critical in a coalition then the sum of the weights of all the players in that coalition (including P_k) is at least q, but the sum of the weights of all the players in the coalition except P_k is less than q. Consequently, if the weights of all the players in that coalition are multiplied by $c > 0$ ($c = 0$ would make no sense), then the sum of the weights of all the players in the coalition (including P_k) is at least cq but the sum of the weights of all the players in the coalition except P_k is less than cq. Therefore P_k is critical in the same coalition in the weighted voting system $[cq: cw_1, cw_2, \ldots, cw_N]$. Since the critical players are the same in both weighted voting systems, the Banzhaf power distributions will be the same.

59. There are $N!$ ways that P_1 (the senior partner) can be the first player in a sequential coalition (this is the number of sequential coalitions consisting of the other N players). When this happens, the senior partner is not pivotal (the second player listed in the sequential coalition is instead). There are $(N+1)! - N! = N!(N+1-1) = N \cdot N!$ ways that P_1 is not the first player in a sequential coalition. In each of these cases, P_1 is the pivotal player. It follows that the Shapley-Shubik power index of P_1 (the senior partner) is $\frac{N \cdot N!}{(N+1)!} = \frac{N \cdot N!}{(N+1) \cdot N!} = \frac{N}{N+1}$. Since the other N players divide the remaining power equally, the junior partners each have a Shapley-Shubik power index of $\frac{1 - \frac{N}{N+1}}{N} = \frac{\frac{N+1}{N+1} - \frac{N}{N+1}}{N} = \frac{1}{N(N+1)}$.

61. You should buy your vote from P_3. The following table explains why.

Buying a vote from	Resulting weighted voting system	Resulting Banzhaf power distribution	Your power
P_1	[8: 5, 4, 2, 2]	$\beta_1 = \frac{4}{12}$; $\beta_2 = \frac{4}{12}$; $\beta_3 = \frac{2}{12}$; $\beta_4 = \frac{2}{12}$	$\frac{2}{12}$
P_2	[8: 6, 3, 2, 2]	$\beta_1 = \frac{7}{10}$; $\beta_2 = \frac{1}{10}$; $\beta_3 = \frac{1}{10}$; $\beta_4 = \frac{1}{10}$	$\frac{1}{10}$
P_3	[8: 6, 4, 1, 2]	$\beta_1 = \frac{6}{10}$; $\beta_2 = \frac{2}{10}$; $\beta_3 = 0$; $\beta_4 = \frac{2}{10}$	$\frac{2}{10}$

63. (a) You should buy your vote from P_2. The following table explains why.

Buying a vote from	Resulting weighted voting system	Resulting Banzhaf power distribution	Your power
P_1	[18: 9, 8, 6, 4, 3]	$\beta_1 = \frac{4}{13}$; $\beta_2 = \frac{3}{13}$; $\beta_3 = \frac{3}{13}$; $\beta_4 = \frac{2}{13}$; $\beta_5 = \frac{1}{13}$	$\frac{1}{13}$
P_2	[18: 10, 7, 6, 4, 3]	$\beta_1 = \frac{9}{25}$; $\beta_2 = \frac{1}{5}$; $\beta_3 = \frac{1}{5}$; $\beta_4 = \frac{3}{25}$; $\beta_5 = \frac{3}{25}$	$\frac{3}{25}$
P_3	[18: 10, 8, 5, 4, 3]	$\beta_1 = \frac{5}{12}$; $\beta_2 = \frac{1}{4}$; $\beta_3 = \frac{1}{6}$; $\beta_4 = \frac{1}{12}$; $\beta_5 = \frac{1}{12}$	$\frac{1}{12}$
P_4	[18: 10, 8, 6, 3, 3]	$\beta_1 = \frac{5}{12}$; $\beta_2 = \frac{1}{4}$; $\beta_3 = \frac{1}{6}$; $\beta_4 = \frac{1}{12}$; $\beta_5 = \frac{1}{12}$	$\frac{1}{12}$

(b) You should buy 2 votes from P_2. The following table explains why.

Buying a vote from	Resulting weighted voting system	Resulting Banzhaf power distribution	Your power
P_1	[18: 8, 8, 6, 4, 4]	$\beta_1 = \frac{7}{27}$; $\beta_2 = \frac{7}{27}$; $\beta_3 = \frac{7}{27}$; $\beta_4 = \frac{1}{9}$; $\beta_5 = \frac{1}{9}$	$\frac{1}{9}$
P_2	[18: 10, 6, 6, 4, 4]	$\beta_1 = \frac{5}{13}$; $\beta_2 = \frac{2}{13}$; $\beta_3 = \frac{2}{13}$; $\beta_4 = \frac{2}{13}$; $\beta_5 = \frac{2}{13}$	$\frac{2}{13}$
P_3	[18: 10, 8, 4, 4, 4]	$\beta_1 = \frac{11}{25}$; $\beta_2 = \frac{1}{5}$; $\beta_3 = \frac{3}{25}$; $\beta_4 = \frac{3}{25}$; $\beta_5 = \frac{3}{25}$	$\frac{3}{25}$
P_4	[18: 10, 8, 6, 2, 4]	$\beta_1 = \frac{9}{25}$; $\beta_2 = \frac{7}{25}$; $\beta_3 = \frac{1}{5}$; $\beta_4 = \frac{1}{25}$; $\beta_5 = \frac{3}{25}$	$\frac{3}{25}$

(c) Buying a single vote from P_2 raises your power from $\frac{1}{25} = 4\%$ to $\frac{3}{25} = 12\%$. Buying a second vote from P_2 raises your power to $\frac{2}{13} \approx 15.4\%$. The increase in power is less with the second vote, but if you value power over money, it might still be worth it to you to buy that second vote.

65. (a) The losing coalitions are $\{P_1\}$, $\{P_2\}$, and $\{P_3\}$. The complements of these coalitions are $\{P_2, P_3\}$, $\{P_1, P_3\}$, and $\{P_1, P_2\}$ respectively, all of which are winning coalitions.

(b) The losing coalitions are $\{P_1\}, \{P_2\}, \{P_3\}, \{P_4\}$, $\{P_2, P_3\}, \{P_2, P_4\}$, and $\{P_3, P_4\}$. The complements of these coalitions are $\{P_2, P_3, P_4\}, \{P_1, P_3, P_4\}, \{P_1, P_2, P_4\}$, $\{P_1, P_2, P_3\}, \{P_1, P_4\}, \{P_1, P_3\}$, and $\{P_1, P_2\}$ respectively, all of which are winning coalitions.

(c) If *P* is a dictator, the losing coalitions are all the coalitions without *P*; the winning coalitions are all the coalitions that include *P*. The complement of any coalition without *P* (losing) is a coalition with *P* (winning).

(d) Take the grand coalition out of the picture for a moment. Of the remaining $2^N - 2$ coalitions, half are losing coalitions and half are winning coalitions, since each losing coalition pairs up with a winning coalition (its complement). Half of $2^N - 2$ is $2^{N-1} - 1$. In addition, we have the grand coalition (always a winning coalition). Thus, the total number of winning coalitions is 2^{N-1}.

67. (a) In each nine-member winning coalition, every member is critical. In each coalition having 10 or more members, only the five permanent members are critical.

(b) At least nine members are needed to form a winning coalition. So, there are 210 + 638 = 848 winning coalitions. Since every member is critical in each nine-member coalition, the nine-member coalitions yield a total of 210 × 9 = 1890 critical players. Since only the permanent members are critical in coalitions having 10 or more members, there are 638 × 5 = 3190 critical players in these coalitions. Thus, the total number of critical players in all winning coalitions is 5080.

(c) Each permanent member is critical in each of the 848 winning coalitions. Thus, the Banzhaf Power Index of a permanent member is 848/5080.

(d) The 5 permanent members together have 5 × 848/5080 = 4240/5080 of the power. The remaining 840/5080 of the power is shared equally among the 10 nonpermanent members, giving each a Banzhaf power index of 84/5080.

(e) In the given weighted voting system, the quota is 39, each permanent member has 7 votes, and each nonpermanent member has 1 vote. The total number of votes is 45 and so if any one of the permanent members does not vote for a measure there would be at most 45 – 7 = 38 votes and the measure would not pass. Thus all permanent members have veto power. On the other hand, all 5 permanent members votes only add up to 35 and so at least 4 nonpermanent members votes are needed for a measure to pass.

Chapter 3

WALKING

A. Shares, Fair Shares, and Fair Divisions

1. (a) Let C = the value of the chocolate half (in Angelina's eyes)
S = the value of the strawberry half (in Angelina's eyes)
It is known that $C = 3S$ (Angelina likes chocolate three times as much as strawberry) and $C + S = \$24$ (the entire cake is worth \$24). Substituting,

$$3S + S = \$24$$
$$4S = \$24$$
$$S = \$6$$

So, the strawberry half is worth \$6.

(b) By (a), the chocolate half is worth $C = 3S = 3(\$6) = \$18$.

(c) $\frac{60°}{180°} \times \$6 = \frac{1}{3} \times \$6 = \$2$

(d) $\frac{45°}{180°} \times \$18 = \frac{1}{4} \times \$18 = \$4.50$

3. (a) Let M = the value of the mushroom half (in Homer's eyes)
P = the value of the strawberry half (in Homer's eyes)
It is known that $P = 4M$ (Homer values pepperoni four times as much as mushroom) and $P + M = \$18$ (the entire pizza is worth \$18). Substituting,

$$4M + M = \$18$$
$$5M = \$18$$
$$M = \$3.60$$

So, the mushroom half is worth \$3.60 to Homer.

(b) By (a), the pepperoni half is worth $P = 4M = 4(\$3.60) = \$14.40$ to Homer.

(c) $\frac{72°}{180°} \times \$3.60 = \$1.44$

(d) $\frac{54°}{180°} \times \$14.40 = \$4.32$

(e) $\frac{30°}{180°} \times \$3.60 + \frac{30°}{180°} \times \$14.40 = \$0.60 + \$2.40 = \$3.00$

5. (a) Let C = the value of the chocolate part (in Karla's eyes)
S = the value of the strawberry part (in Karla's eyes)
V = the value of the vanilla part (in Karla's eyes)
It is known that $S = 2V$ (Karla likes strawberry twice as much as vanilla), $C = 3V$ (Karla likes chocolate three times as much as vanilla), and $C + S + V = \$30$ (the entire cake is worth \$30). Substituting,

$$3V + 2V + V = \$30$$
$$6V = \$30$$
$$V = \$5$$
$$S = 2\times(\$5) = \$10$$
$$C = 3\times(\$5) = \$15$$

That is, the vanilla part is worth \$5, the strawberry part \$10, and the chocolate part \$15 in Karla's eyes. The value of each slice is then found as follows.

$s_1 : \frac{60°}{120°}\times\$5 = \$2.50$; $s_2 : \frac{30°}{120°}\times\$5 + \frac{30°}{120°}\times\$10 = \$1.25 + \$2.50 = \$3.75$;

$s_3 : \frac{60°}{120°}\times\$10 = \$5.00$; $s_4 : \frac{30°}{120°}\times\$10 + \frac{30°}{120°}\times\$15 = \$2.50 + \$3.75 = \$6.25$;

$s_5 : \frac{60°}{120°}\times\$15 = \$7.50$; $s_6 : \frac{30°}{120°}\times\$15 + \frac{30°}{120°}\times\$5 = \$3.75 + \$1.25 = \$5.00$

(b) A fair share would need to be worth at least \$30/6 = \$5.00. So, slices s_3, s_4, s_5, and s_6 are each fair shares.

7. (a) $\frac{1}{3}\times(\$3.00 + \$5.00 + \$4.00) = \frac{1}{3}\times(\$12.00) = \$4.00$

Any slice worth at least \$4.00 is a fair share to Ana. So, slices s_2 and s_3 are fair shares to Ana.

(b) $\frac{1}{3}\times(\$4.00 + \$4.50 + \$6.50) = \frac{1}{3}\times(\$15.00) = \$5.00$

Any slice worth at least \$5.00 is a fair share to Ben. So, slice s_3 is a fair share to Ben.

(c) $\frac{1}{3}\times(\$4.50 + \$4.50 + \$4.50) = \frac{1}{3}\times(\$13.50) = \$4.50$

Any slice worth at least \$4.50 is a fair share to Cara. So, slices s_1, s_2, and s_3 are fair shares to Cara.

(d) Ben must receive s_3; So, Ana must receive s_2 and then Cara is left with s_1.

9. (a) Completing the table may be helpful.

	s_1	s_2	s_3
Alex	30%	40%	30%
Betty	31%	35%	34%
Cindy	30%	35%	35%

Alex considers only one share, s_2, to be worth 1/3 or more of the cake.

(b) Betty considers two shares (s_2, s_3) to be each worth 1/3 or more of the entire cake.

(c) s_2, s_3

(d) It is not possible. No player considers s_1 to be a fair share.

11. (a) Since Abe values a fair share at $(\$3.00 + \$5.00 + \$5.00 + \$2.00)/4 = \$15.00/4 = \$3.75$, he considers slices s_2 and s_3 to be fair shares.

(b) s_1, s_2, s_3, s_4; Betty considers each share to be worth the same amount and hence fair.

(c) Since Cory values a fair share at $(\$4.00+\$3.50+\$2.00+\$2.50)/4=\$12.00/4=\$3.00$, he considers slices s_1 and s_2 to be fair shares.

(d) Since Dana values a fair share at $(\$2.75+\$2.40+\$2.45+\$2.40)/4=\$10.00/4=\$2.50$, she considers only slice s_1 to be fair.

(e) Dana must receive s_1. Then, Cory must receive s_2. After that, Abe must receive s_3 leaving Betty with s_4.

13. (a) Let x denote the value of slices s_2 and s_3 to Adams. Then, $(x+\$40,000)+x+x+(x+\$60,000)=\$400,000$ and so x = \$75,000. Since a fair share is worth \$100,000, Adams considers slices s_1 (\$115,000) and s_4 (\$135,000) to be fair shares.

(b) Let x denote the value of slice s_3 to Benson. Then, since s_4 is worth \$8000 more than that and the sum of these two is worth $(0.40)\$400,000=\$160,000$, it follows that $x+(x+\$8,000)=\$160,000$ and so x = \$76,000. So, s_3 is worth \$76,000 and s_4 is worth \$84,000.
Now let y denote the value of slice s_2 to Benson. The value of s_1 is then $y+\$40,000$. So, $(y+\$40,000)+y+\$76,000+\$84,000=\$400,000$ and $y=\$100,000$. Since a fair share is \$100,000, Benson considers slices s_1 (\$140,000) and s_2 (\$100,000) to be fair shares.

(c) Let x denote the value of slice s_4 to Cagle. Then, s_3 is worth $2x$ and s_1 is worth x + \$20,000. Since s_2 is worth \$40,000 less than s_1, the value of s_2 is x - \$20,000. Hence, $(x+\$20,000)+(x-\$20,000)+2x+x=\$400,000$ and x = \$80,000. Since a fair share is \$100,000, Cagle considers slices s_1 (\$100,000) and s_3 (\$160,000) to be fair shares.

(d) Let x denote the value of slices s_2 and s_3 to Duncan. Then, $(x+\$4,000)+x+x=(0.70)(\$400,000)$ = \$280,000 so that x = \$92,000. Since a fair share is \$100,000, Duncan only considers slice s_4 (\$120,000) to be a fair share.

(e) Duncan must receive s_4. Then, Adams must receive s_1. After that, Benson must receive s_2 leaving Cagle with s_3.

B. The Divider–Chooser Method

15. (a) Since Jared likes meatball subs three times as much as vegetarian subs, he values each of the pieces [0,6], [6,8], [8,10], and [10,12] the same. So, with one cut Jared would divide the sandwich into parts s_1 : [0,8] and s_2 :[8,12]. That is, one piece containing the entire vegetarian half and one-third of the meatball half (s_1) and one piece consisting of two-thirds of the meatball half (s_2).

(b) Karla would clearly choose piece s_1 since she only values vegetarian. This piece contains the entire value of the sandwich to Karla (\$8.00).

17. (a) Let x denote the value that Martha places on one inch of turkey (in dollars). Then, Martha places that same value on one inch of roast beef. However, she places a value of $2x$ on each inch of ham. Since the total value of the sandwich is \$9, we have $8(2x)+12(x)+8(x)=\$9$. That is, $36x$ = \$9 or x = \$0.25.

Each inch of ham is worth $0.50 to Martha and each inch of turkey and roast beef is worth $0.25. To have two pieces each worth $4.50, she would cut the sandwich into pieces s_1 : [0,10] and s_2 :[10,28].

(b) Let *y* denote the value that Nick places on one inch of turkey (in dollars). Then, Nick places that same value on one inch of ham. However, he places a value of 2*y* on each inch of roast beef. Since the total value of the sandwich is $9, we have $8(y)+12(y)+8(2y)=\$9$. That is, 36*y* = $9 or *y* = $0.25. Each inch of roast beef is worth $0.50 to Nick and each inch of turkey and ham is worth $0.25. Nick would clearly take piece s_2 :[10,28] which, in his eyes, has a value of $10(\$0.25)+8(\$0.50)=\$6.50$.

19. (a) Yes; David values pepperoni and sausage twice as much as mushroom. Paula would choose s_2 since she hates pepperoni (s_2 contains ¾ of the value of the entire pizza in her eyes).

(b) No; David values s_2 more than he values s_1. In fact, he values it 1.5 times more. Such a division is not rational since there is a risk that Paula, whom he knows nothing about, would choose that piece.

(c) Yes; Based on David's value system, the pepperoni and sausage parts are each worth twice as much as the mushroom part. So, if the entire pizza is worth $5, then, s_1 is worth $\frac{90^\circ}{120^\circ}(\$2)+\frac{60^\circ}{120^\circ}(\$2)=\$2.50$, or half that amount. Paula would choose the piece with the least pepperoni (piece s_2).

C. The Lone-Divider Method

21. (a) *D* must be given s_1. But C_1 and C_2 are happy with either of the other shares. So, the following are two fair divisions of the land.

$C_1 : s_2,\ C_2 : s_3, D : s_1$;
$C_1 : s_3,\ C_2 : s_2, D : s_1$

(b) If C_1 is assigned s_2, then C_2 can be assigned either of the other pieces. However, if C_1 is assigned s_3, then C_2 must be assigned s_1. So, the following are three fair divisions.

$C_1 : s_2,\ C_2 : s_1, D : s_3$;
$C_1 : s_2,\ C_2 : s_3, D : s_1$;
$C_1 : s_3,\ C_2 : s_1, D : s_2$

23. (a) First, C_1 must receive s_2. Then, C_3 must receive s_3 (the only other share C_3 considers fair). So, C_2 receives s_1 and *D* receives s_4.

(b) If C_1 receives s_2, then C_3 must receive s_1. But then C_2 must receives s_3 leaving *D* with s_4. If, on the other hand, C_1 receives s_3, then C_2 must receive s_1. But then C_3 must receives s_2 leaving *D* with s_4. Hence, two fair divisions are as follows.

$C_1 : s_2,\ C_2 : s_3, C_3 : s_1, D : s_4$;
$C_1 : s_3,\ C_2 : s_1, C_3 : s_2, D : s_4$

(c) C_1 must receive s_2. If C_2 receives s_1, then C_3 must receive s_4 leaving *D* with s_3. On the other hand, if C_2 receives s_3, then C_3 could receive either s_1 or s_4 leaving *D* with the remaining piece. Three fair divisions follow below.

$C_1 : s_2,\ C_2 : s_1, C_3 : s_4, D : s_3$;
$C_1 : s_2,\ C_2 : s_3, C_3 : s_1, D : s_4$;
$C_1 : s_2,\ C_2 : s_3, C_3 : s_4, D : s_1$

25. (a) Fair; none of the bidders considered s_3 as being worth ¼ of the land. So, these three shares must be worth at least ¾ of the land in each of their eyes. If they each wind up with at least 1/3 of the newly combined piece, that would constitute at least 1/3 of 3/4 (i.e. ¼) of the value of the original plot of land.

(b) Not fair; $\{s_2, s_3, s_4\}$ may be worth less than ¾ of the value of the original plot of land to C_2 and C_3. Even if they wind up with 1/3 of the newly combined piece, it may not be a fair share.

(c) Fair; see (a).

(d) Not fair; $\{s_1, s_4\}$ may be worth less than ½ of the value of the original plot of land to C_2 and C_3. That is, s_2 may be of a lot of value to C_2 and C_3.

27. (a) $C_1 : s_2,\ C_2 : s_4, C_3 : s_5, C_4 : s_3, D : s_1$;
$C_1 : s_4,\ C_2 : s_2, C_3 : s_5, C_4 : s_3, D : s_1$;
If C_1 is to receive s_2, then C_2 must receive s_4 (and conversely). So, C_4 must receive s_3. It follows that C_3 must receive s_5. This leaves D with s_1.

(b) $C_1 : s_2,\ C_2 : s_4, C_3 : s_5, C_4 : s_3, D : s_1$;
If C_1 is to receive s_2, then C_2 must receive s_4. So, C_4 must receive s_3. It follows that C_3 must receive s_5. This leaves D with s_1.

29. (a) The Divider was Gong since that is the only player that could possibly value each piece equally.

(b) To determine the bids placed, each player's bids should add up to $480,000. So, for example, Egan would bid $480,000 - $80,000 - $85,000 - $195,000 = $120,000 on parcel s_3. Further, one player must bid the same on each parcel (the divider). The following table shows the value of the four parcels in the eyes of each partner.

	s_1	s_2	s_3	s_4
Egan	$80,000	$85,000	$120,000	$195,000
Fine	$125,000	$100,000	$135,000	$120,000
Gong	$120,000	$120,000	$120,000	$120,000
Hart	$95,000	$100,000	$175,000	$110,000

For a player to bid on a parcel, it would need to be worth at least $480,000/4 = $120,000. Based on this table, the choosers would make the following declarations:
Egan: $\{s_3, s_4\}$; Fine: $\{s_1, s_3, s_4\}$; Hart: $\{s_3\}$.

(c) One possible fair division of the land is

Egan	**Fine**	**Gong**	**Hart**
s_4	s_1	s_2	s_3

In fact, this is the only possible division of the land since Hart must receive s_3, so that Egan in turn must receive s_4. It follows that Fine must receive s_1 so that Gong (the divider) winds up with s_2.

31. (a) $s_1 : [0,4]$; $s_2 : [4, 8]$; $s_3 : [8, 12]$

(b) Karla values [0,2], [2,4], and [4,6] equally. She places no value on the segment [6,12]. So, Karla would consider any piece that contains [0,2] or [2,4] or [4,6] to be a fair share. It follows that Karla views s_1 and s_2 as fair shares.

(c) Suppose that Lori places a value of $1 (or 1 cent, or one peso, or whatever) on one inch of vegetarian sub. To her then the entire sub is worth $6 + $12 = $18. Also, she values s_1 at $4, s_2 at $6, and s_3 at $8. Since a fair share must be worth $6 to Lori, it follows that s_2 and s_3 are fair shares.

(d) Three possible fair divisions are
Jared: s_2, Karla: s_1, Lori: s_3;
Jared: s_1, Karla: s_2, Lori: s_3;
Jared: s_3, Karla: s_1, Lori: s_2

D. The Lone-Chooser Method

33. (a) Angela sees s_1 as being worth $18. In her second division, she will create three $6 pieces.

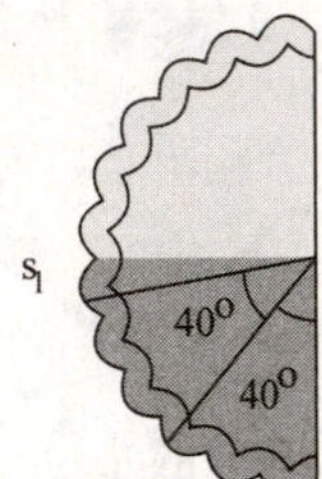

Since the strawberry quarter at the lower left is worth $13.50 to Angela, the size of the central angle in one pie slice she will cut will be determined by

$\frac{x}{90^\circ}(\$13.50) = \$6$ so that $x = 40^\circ$.

In fact, Angela cuts two 40° pieces from the strawberry, and the remaining strawberry plus the vanilla makes up the third piece. The pieces can be described as 40° strawberry, 40° strawberry, and 90° vanilla-10° strawberry.

(b) Boris sees s_2 as being worth $15. In his second division, he will create three $5 pieces.

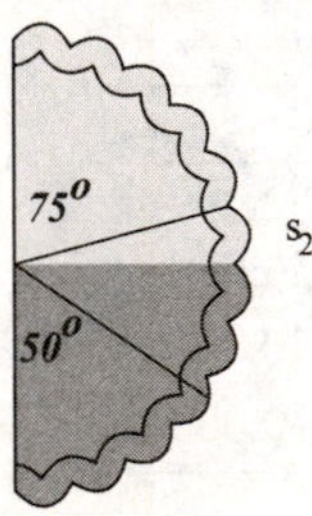

Since the strawberry quarter at the lower right is worth $9.00 to Boris, the size of the central angle in one pie slice he will cut will be determined by $\frac{x}{90^\circ}(\$9.00) = \$5.00$ so that $x = 50^\circ$. So Boris cuts one 50° piece from the strawberry part of the cake. Also, since the vanilla quarter at the upper right is worth $6.00 to Boris, we see that $\frac{x}{90^\circ}(\$6.00) = \$5.00$ tells us that $x = 75^\circ$ is the size of the central angle in a second pie slice (vanilla this time) Boris makes. The remaining 15° piece of vanilla plus the 40° piece of strawberry makes up the third piece. The pieces can be described as 75° vanilla, 15° vanilla-40° strawberry, 50° strawberry.

(c) Since Carlos values vanilla twice as much as strawberry, Carlos would clearly take the vanilla part of Angela's division (the 90° vanilla-10° strawberry wedge). He would also want to take the most vanilla that he could from Boris (the 75° vanilla wedge). One possible fair division is Angela: 80° strawberry, Boris: 15° vanilla-90° strawberry, and Carlos: 165° vanilla-10° strawberry.

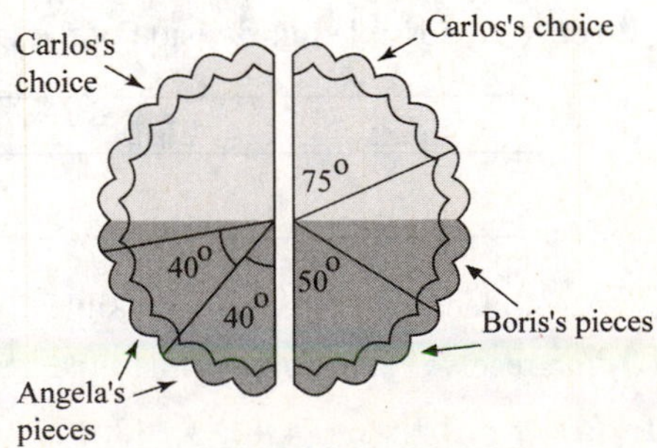

(d) Since she receives two of her \$6.00 pieces, the value of Angela's final share (in Angela's eyes) is \$12.00. Similarly, the value of Boris' final share (in Boris' eyes) is \$10.00. The value of Carlos' final share (in Carlos' eyes) is $\frac{90^\circ}{90^\circ}(\$12)+\frac{10^\circ}{90^\circ}(\$6)+\frac{75^\circ}{90^\circ}(\$12)\approx\$22.67$.

35. First, note that the value of s_1 (in Angela's eyes) is $\frac{120^\circ}{180^\circ}\times\$27=\$18$. She values s_2 at \$18 too.

(a) Boris chooses s_2 because he views it as being worth

$\$12+\frac{60^\circ}{180^\circ}\times\$18=\$12+\$6=\$18$ (s_1 is only worth \$12.00 to him).

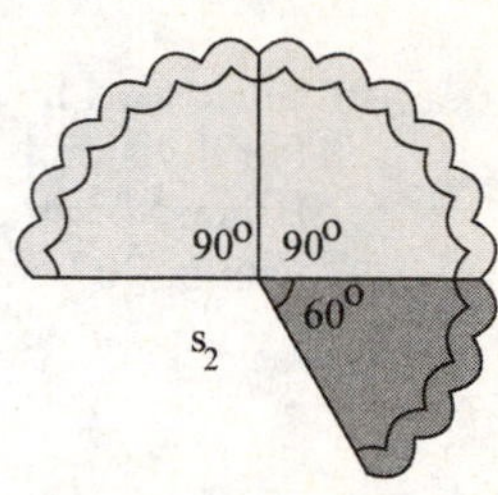

Boris views s_2 as being worth \$18 and divides it into three pieces each worth \$6.00. Thus, one possible second division of s_2 by Boris is 90° vanilla, 90° vanilla, 60° strawberry.

(b) Angela divides s_1 evenly (40° strawberry, 40° strawberry, 40° strawberry).

(c) Carlos will select any one of Angela's pieces since they are identical in value to him. He will also select one of Boris's vanilla wedges since he values vanilla twice as much as he values strawberry.

One possible fair division is Angela: 80° strawberry, Boris: 90° vanilla-60° strawberry, and Carlos: 90° vanilla-40° strawberry.

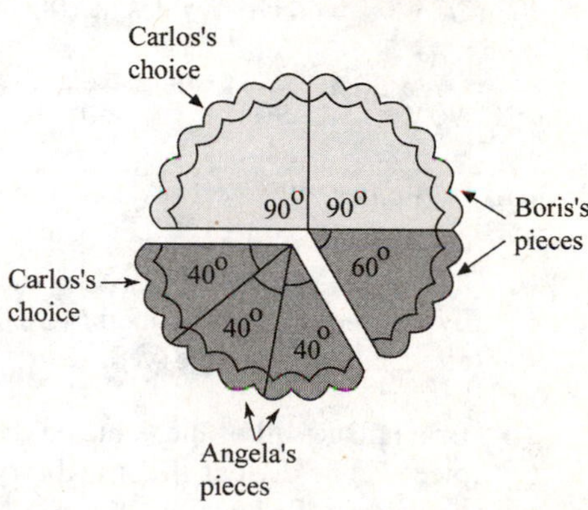

(d) Angela thinks her share is worth $\frac{80^\circ}{180^\circ}\times\$27=\$12.00$.

Boris thinks his share is worth $\frac{90^\circ}{180^\circ}\times\$12+\frac{60^\circ}{180^\circ}\times\$18=\$12.00$.

Carlos thinks his share is worth $\frac{40^\circ}{180^\circ}\times\$12+\frac{90^\circ}{180^\circ}\times\$24=\$14.67$.

37. **(a)** After Arthur makes the first cut, Brian chooses s_1. To him, it is worth 100% of the cake. One possible second division by Brian is to divide the piece evenly (60° chocolate, 30° chocolate-30° strawberry, 60° strawberry).

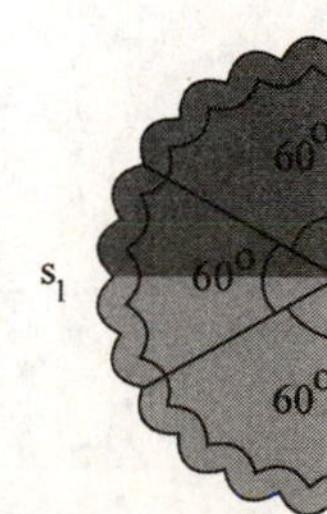

(b) Arthur places all of the value on the orange half of s_2, so he divides the orange evenly. The second division by Arthur can be described as 30° orange, 30° orange, 30° orange-90° vanilla.

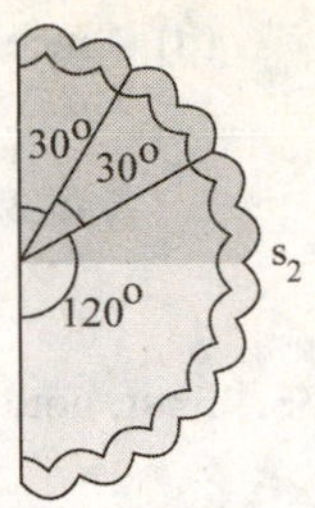

(c) Since Carl likes chocolate and vanilla, one possible fair division is Arthur: 60° orange, Brian: 90° strawberry-30° chocolate, and Carl: 90° vanilla-30° orange-60° chocolate.

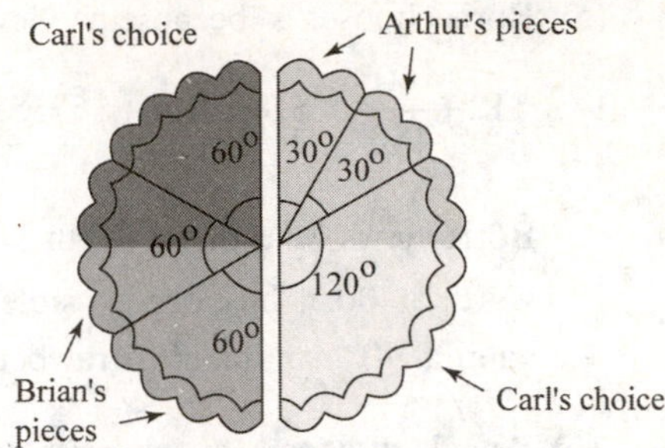

(d) Arthur thinks his share is worth $33\frac{1}{3}$%. Brian thinks his share is worth $66\frac{2}{3}$% (remember, he is getting 2/3 of s_1 which he values at 100%). Carl thinks his share is worth

$$\frac{60°}{90°}\times 50\% + \frac{90°}{90°}\times 50\% + \frac{30°}{90°}\times 0\% = 83\frac{1}{3}\%.$$

39. **(a)** Carl picks either s_1 or s_2 (we assume he picks s_2). It is worth (to him) 50% of the value of the whole cake. Carl places all of the value on the chocolate half of the piece, so he divides the chocolate evenly. So, a second division by Carl could be described as a 30° chocolate, 30° chocolate, and 30° chocolate-90° orange wedge.

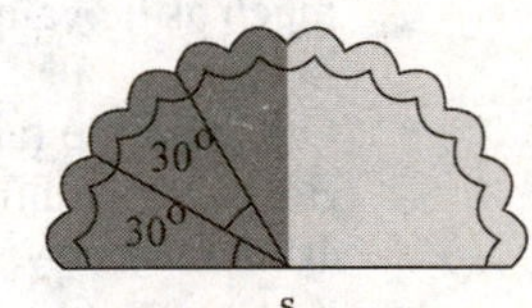

(b) Brian places all of the value on the strawberry half of the piece, so he divides the strawberry evenly. A second division by Brian could be described as 30° strawberry, 30° strawberry, 30° strawberry-90° vanilla.

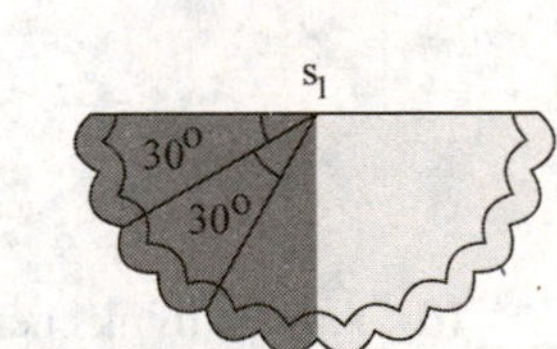

(c) One possible fair division is Arthur: 30° chocolate-90° orange-30° strawberry, Brian: 60° strawberry-90° vanilla, and Carl: 60° chocolate.

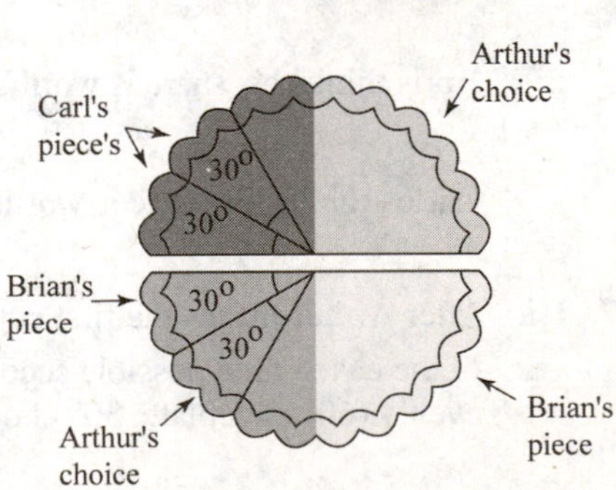

(d) Arthur thinks his share is worth $\frac{90°}{90°}\times 50\% + \frac{30°}{90°}\times 50\% = 66\frac{2}{3}\%$. Brian thinks his share is worth $33\frac{1}{3}$%. Carl thinks his share is worth $33\frac{1}{3}$%.

41. (a) Ignoring the meatball part (having no value to her), Karla would cut the sandwich right down the middle of the vegetarian part. That is, s_1 : [0,3], s_2 :[3,12].

(b) Since Jared likes meatball and vegetarian the same, he will pick the larger half (by size), namely s_2 . He will then divide his choice (s_2 :[3,12]) equally as J_1 :[3,6], J_2 :[6,9], J_3 :[9,12].

(c) Karla's subdivision is just as simple as Jared's – she likes all parts of s_1 equally. So her division can be described as K_1 :[0,1], K_2 :[1,2], K_3 :[2,3].

(d) Since Lori likes meatball twice as much as vegetarian, she will select any of Karla's identical pieces (say, K_3 :[2,3] is selected). She will also select either J_2 or J_3 from Jared (say, J_3 :[9,12]). So, one fair division would be described by Jared: [3,9], Karla: [0,2], Lori: [2,3] and [9,12]. In this case, Jared ends up with half of the sandwich (in size) worth 50% of the value of the whole sandwich to him. Karla ends up with 1/3 of the vegetarian half of the sandwich worth $33\frac{1}{3}\%$ of the value of the whole sandwich to her. Lori values the vegetarian half at 1/3 and the meatball part at 2/3 so that her piece is valued at $\frac{1}{6}\times\frac{1}{3}+\frac{1}{2}\times\frac{2}{3}=\frac{7}{18}=38\frac{8}{9}\%$.

E. The Last-Diminisher Method

43. (a) P_2 and P_4 are both diminishers since they value the piece when it is their turn as worth more than \$30/5 = \$6.

(b) P_4 values the piece as being worth \$6.50 when it is their turn to play so they diminish its value to \$6.00 when they make a claim. As the last diminisher in that round, P_4 gets their fair share at the end of round 1. The share, to her, is worth \$6.00.

(c) P_1 makes the first cut to start round 2 (again!). The remaining players consist of everyone except for P_4 . They divide what remains, the R-piece, into quarters. The value of that piece is worth at least \$30 - \$6 = \$24. That is, the value of the R-piece to each of the remaining players is $\geq$\$24.

45. (a) P_3 and P_4 are both diminishers since they value the piece when it is their turn as worth at least \$300,000/4 = \$75,000.

(b) P_4 values the piece as being worth \$80,000 when it is their turn to play. However, as the last player in that round they choose to diminish the piece by 0%. P_4 gets their fair share at the end of round 1 worth, to her, \$80,000.

(c) P_1 makes the first cut to start round 2 (again!). The remaining three players divide what remains, the R-piece, into thirds. The value of that piece is worth at least \$300,000 - \$75,000 = \$225,000. That is, the value of the R-piece to each of the remaining players is $\geq$\$225,000. Remember, this is not in the eyes of player P_4 .

47. (a) P_9 ; she is the last diminisher in round 1.

(b) P_1

(c) P_5 ; he is the last (and only) diminisher in round 2.

(d) P_1 ; she is the only claimant.

(e) P_2

(f) 11; One fair share is given away in each round. However, the last round is not necessary as the last piece will automatically go to the last player.

49. (a) Boris, playing second, will diminish the claim made by Carlos. Since he values the strawberry half of the cake at \$18, he will diminish it to a 100° strawberry wedge (worth \$10 to him which is 1/3 of what he values the entire cake). Angela, the last player, will claim the 100° strawberry wedge and diminish it 0%. In the end, Angela gets a 100° strawberry wedge.

(b) Carlos

(c) At the beginning of round two, the cake is worth $\$24+\frac{80^\circ}{180^\circ}(\$12)\approx\$29.33$ to Carlos. A fair share would be worth \$14.67 to him. So, Carlos starts round two by cutting a wedge that is $\frac{\$14.67}{\$24}\approx 61.111\%$ of the vanilla half of the cake. That is, he cuts a 110° vanilla wedge. Boris passes (he values vanilla less than Carlos). Carlos is the last diminisher and receives a 110° vanilla wedge.

(d) Angela receives a 100° strawberry wedge worth $\frac{100^\circ}{180^\circ}(\$27)=\$15$ to her. This is worth, in her eyes, $\frac{15}{36}=\frac{5}{12}$ of the whole cake.

Boris receives an 80° strawberry wedge worth $\frac{80^\circ}{180^\circ}(\$18)=\$8$ and a 70° vanilla wedge worth $\frac{70^\circ}{180^\circ}(\$12)=\$4.67$ to him. Both together are worth, in his eyes, $\frac{12.67}{30}=\frac{19}{45}$ of the whole cake.

Carlos receives a 110° vanilla wedge worth $\frac{110^\circ}{180^\circ}(\$24)=\$14.67$ to him. To him, this is worth $\frac{\$14.67}{\$36}=\frac{11}{27}$ of the whole cake.

51. (a) Let M denote the value of an inch of the meatball part of the sub to Lori. Then, an inch of the vegetarian part of the sub is valued at $3M$. The entire sub is worth $6M+6(3M)=24M$. So an inch of the meatball part of the sub is worth $\frac{1}{24}$ of the whole sandwich. An inch of the vegetarian part is worth $\frac{3}{24}$ of the whole. Solving $\frac{3}{24}x=\frac{1}{3}$ for x give $x=2\frac{2}{3}$.

(b) Karla values [0,2], [2,4], and [4,12] equally. So, she will diminish the C-piece to [0,2]. Jared sees [0,2] as only being worth half of a fair share to him (he considers [0,4], [4,8], and [8,12] as fair shares). So he passes and Karla is the last diminisher in round 1. She receives [0,2] (i.e. two inches of the vegetarian part of the sandwich).

(c) Lori is first to claim a C-piece on R (which is now [2,12]) in round 2. Since each inch of vegetarian sub is worth three times as much as each inch of meatball sub, Lori values [2,5] and [5,12] equally. Suppose she claims [5,12]. Then, Jared will diminish it 0% and be the last diminisher at the end of round 2. The final fair division of the sandwich would be then described as Jared: [5,12] , Karla: [0,2], and Lori: [2,5].

F. The Method of Sealed Bids

53. (a) Ana and Belle's fair share is \$300. Chloe's fair share is \$400.

Item	**Ana**	**Belle**	**Chloe**
Dresser	150	**300**	275
Desk	**180**	150	165
Vanity	170	200	**260**
Tapestry	400	250	**500**
Total Bids	900	900	1200
Fair share	300	300	400

(b) In the first settlement, Ana receives the desk and $120 in cash; Belle receives the dresser; Chloe gets the vanity and the tapestry and pays $360.

	Ana	**Belle**	**Chloe**
Total Bids	900	900	1200
Fair share	300	300	400
Value of items received	180	300	760
Prelim cash settlement	120	0	–360

(c) $240; $360 was paid in and $120 was paid out in the first settlement.

(d) Ana, Belle, and Chloe split the $240 surplus cash three ways. This adds $80 cash to the first settlement. In the final settlement, Ana gets the desk and receives $200 in cash; Belle gets the dresser and receives $80; Chloe gets the vanity and the tapestry and pays $280.

Item	**Ana**	**Belle**	**Chloe**
Prelim cash settlement	120	0	–360
Share of surplus	80	80	80
Final cash settlement	200	80	–280

55. **(a)** Each player's fair share is the sum of their bids divided by 5.

Item	***A***	***B***	***C***	***D***	***E***
Item 1	$352	$295	**$395**	$368	$324
Item 2	$98	$102	$98	$95	**$105**
Item 3	$460	$449	**$510**	$501	$476
Item 4	**$852**	$825	$832	$817	$843
Item 5	**$513**	$501	$505	$505	$491
Item 6	$725	$738	$750	$744	**$761**
Total Bids	$3000	$2910	$3090	$3030	$3000
Fair Share	$600	$582	$618	$606	$600

(b) After the first settlement, *A* ends up with items 4 and 5 and pays $765, *B* gets $582, *C* receives items 1 and 3 and pays $287, *D* gets $606, and *E* receives items 2 and 6 and pays $266.

	A	*B*	*C*	*D*	*E*
Total Bids	3000	2910	3090	3030	3000
Fair Share	600	582	618	606	600
Value of items rec'd	1365	0	905	0	866
Prelim. cash	–765	582	–287	606	–266

(c) There is a surplus of \$765 - \$582 + \$287 - \$606 + \$266 = \$130.

(d) *A*, *B*, *C*, *D*, and *E* split the \$130 surplus 5 ways (\$26 each). In the final settlement, *A* ends up with items 4 and 5 and pays \$739; *B* ends up with \$608; *C* ends up with items 1 and 3 and pays \$261; *D* ends up with \$632; *E* ends up with items 2 and 6 and pays \$240.

57. Suppose that Angelina bid \$$x$ on the laptop. Then, she values the entire estate at x+\$2900. Brad, on the other hand, values the entire estate at \$4640. The rest of the story is given in the table below. Note: Knowing Brad's final cash settlement of paying \$355 determines that his share of the surplus is \$460 - \$355 = \$105.

	Angelina	**Brad**
Fair Share	(x+\$2900)/2	\$2320
Value of items rec'd	x+\$300	\$2780
Prelim. cash	(x+\$2900)/2 - (x+\$300)	-\$460
Share of surplus	\$105	\$105
Final cash	\$355	-\$355

To determine the value of x, it must be that $\frac{\$2300-x}{2}+\$105=\$355$. This leads to $\$2300-x=\$500$ and $x=$ \$1800. Angelina bid \$1800 on the laptop.

59. Anne gets \$75,000 and Chia gets \$80,000.

Item	**Anne**	**Bette**	**Chia**	
Partnership	\$210,000	\$240,000	\$225,000	
Total Bids	\$210,000	\$240,000	\$225,000	
Fair Share	\$70,000	\$80,000	\$75,000	**Total**
Value of items received	0	\$240,000	0	**Surplus**
Prelim. cash settlement	\$70,000	–\$160,000	\$75,000	**\$15,000**
Share of surplus	\$5,000	\$5,000	\$5,000	
Final cash settlement	\$75,000	–\$155,000	\$80,000	

G. The Method of Markers

61. (a) Using the first first marker, *B* gets items 1, 2, 3; Then, using the first second marker, *C* gets items 5, 6, 7; Finally, *A* gets items 10, 11, 12, 13.

(b) *B* gets items 1, 2, 3.

(c) *C* gets items 5, 6, 7.

(d) Items 4, 8, and 9 are left over.

63. **(a)** Using the first first marker, *A* gets items 1, 2; Then, using the first second marker, *C* gets items 4, 5, 6, 7; Finally, *B* gets items 10, 11, 12.

(b) *B* gets items 10, 11, 12

(c) *C* gets items 4, 5, 6, 7

(d) Items 3, 8, and 9 are left over.

65. **(a)** First *C* gets items 1, 2, 3; Then, *E* gets items 5, 6, 7, 8; Then, *D* gets items 11, 12, 13; Then, *B* gets items 15, 16, 17; Finally, *A* gets items 19, 20.

(b) Items 4, 9, 10, 14, and 18 are left over.

67. **(a)** Quintin thinks the total value is $3\times\$12+6\times\$7+6\times\$4+3\times\$6=\$120$, so to him a fair share is worth \$30. Ramon thinks the total value is $3\times\$9+6\times\$5+6\times\$5+3\times\$11=\$120$, so to him a fair share is worth \$30. Stephone thinks the total value is $3\times\$8+6\times\$7+6\times\$6+3\times\$14=\$144$, so to him a fair share is worth \$36. Tim thinks the total value is $3\times\$5+6\times\$4+6\times\$4+3\times\$7=\$84$, so to him a fair share is worth \$21. They would place their markers as shown below.

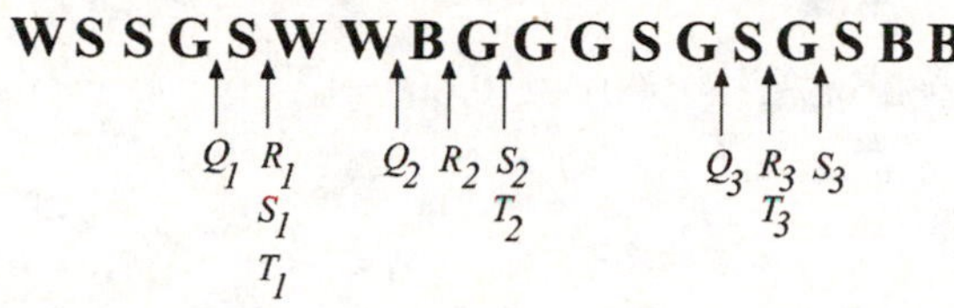

(b)

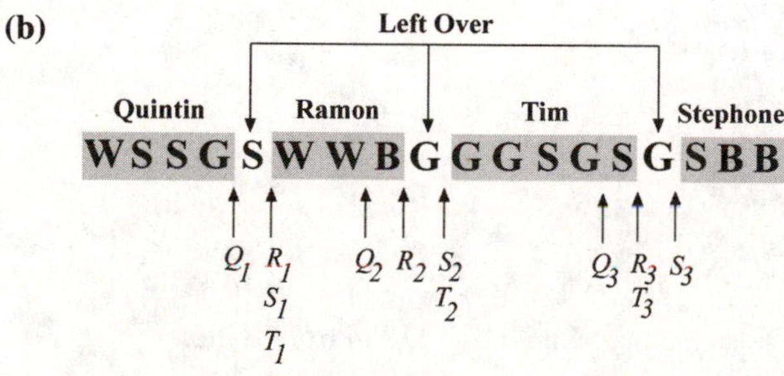

69. **(a)** Ana places her markers between the different types of candy bars. Belle places her markers between each of the Nestle Crunch bars. Chloe places her markers thinking that each Snickers and Nestle Crunch bar is worth \$1 and each Reese's is worth \$2. Since Chloe would then value the bars to be worth \$12 in total, she would place markers after each \$4 worth of bars.

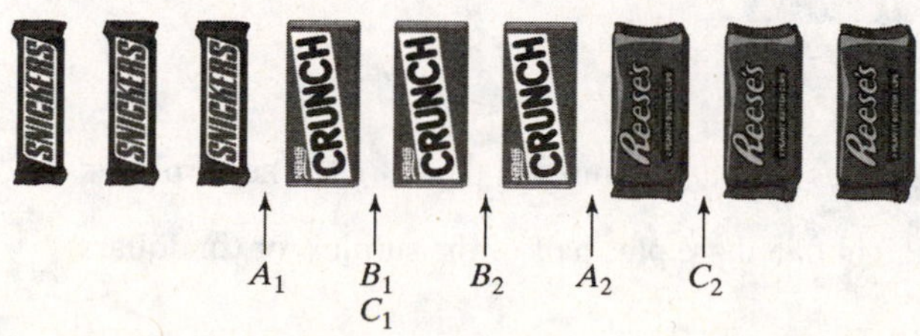

(b) Ana gets three Snickers, Belle gets one Nestle Crunch, Chloe gets two Reese's Peanut Butter Cups. Two Nestle Crunch bars and one Reese's Peanut Butter Cup are left over.

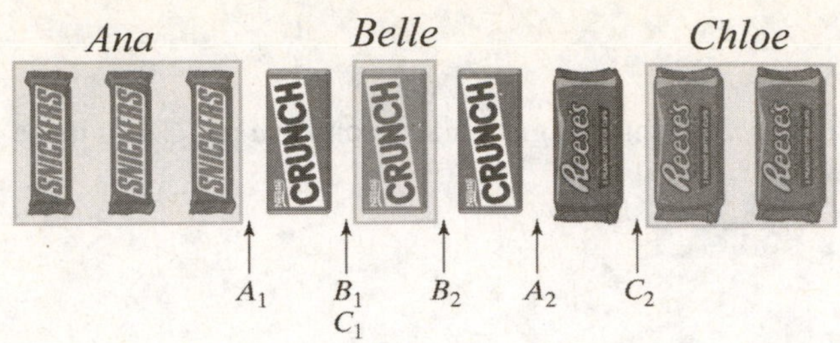

(c) Belle would select one of the Nestle Crunch bars, Chloe would then select the Reese's Peanut Butter Cup, and then Ana would be left with the last Nestle Crunch bar.

JOGGING

71. **(a)** The total area is 30,000 m^2 and the area of C is only $\frac{100(50+110)}{2} = 8000\, m^2$. Since P_2 and P_3 value the land uniformly, each thinks that a fair share must have an area of at least 10,000 m^2.

(b) The cut parallel to Park Place which divides the parcel in half is illustrated below. The cut is made x meters from the bottom. We know that $\frac{y}{x} = \frac{60}{100}$ so that $y = \frac{3}{5}x$. The bottom trapezoid is to have area of 11,000 m^2. So, $\frac{(190+(3/5)x)+190}{2}x = 11,000$ or $3x^2 + 1900x - 110,000 = 0$. By the quadratic formula, the value of x is $\frac{-950+50\sqrt{493}}{3} \approx 53.4$ m.

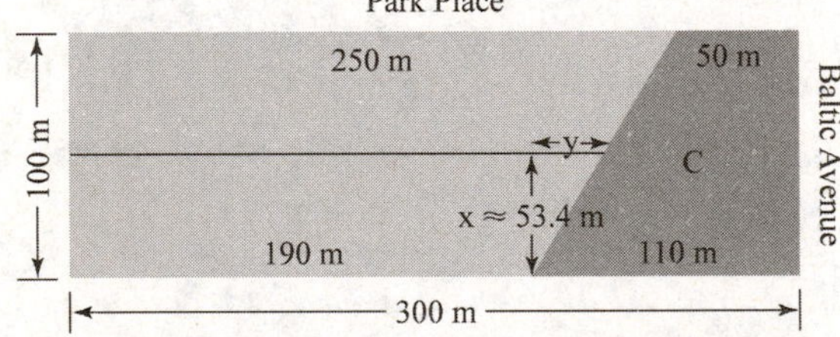

(c) The cut parallel to Baltic Avenue which divides the parcel in half is 110 m from Baltic.

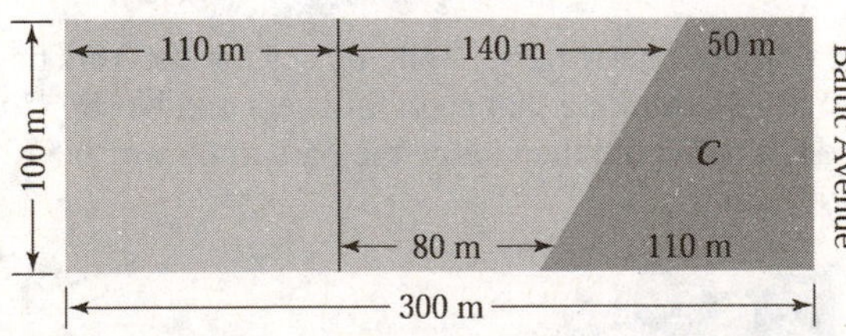

73. In the first settlement A receives \$$x$/2; B receives the partnership and pays \$$y$/2. The surplus is $\frac{y}{2} - \frac{x}{2} = \frac{y-x}{2}$ dollars. B must pay A's original fair share plus half of the surplus, or (in dollars)

$$\frac{x}{2} + \frac{1}{2}\frac{y-x}{2} = \frac{x}{2} + \frac{y-x}{4} = \frac{x}{2} + \frac{y}{4} - \frac{x}{4} = \frac{x}{4} + \frac{y}{4} = \frac{x+y}{4}.$$

75. B receives the strawberry part. C will receive 5/6 of the chocolate part. A receives the vanilla part and 1/6 of the chocolate part.

77. **(a)** Dandy bids $\{s_1, s_2, s_3, s_4\}$, Burly bids $\{s_1\}$, Curly bids $\{s_1, s_2\}$, and Greedy bids $\{s_1, s_2, s_3\}$. So the final division is Dandy: s_4; Burly: s_1; Curly: s_2; Greedy: s_3.

(b) If Greedy only bids on s_1, then the bid lists are as follows: Dandy: $\{s_1, s_2, s_3, s_4\}$; Burly: $\{s_1\}$; Curly: $\{s_1, s_2\}$; Greedy: $\{s_1\}$. Suppose that Curly is given s_2 and Dandy is given s_3. The parcels s_1 and s_4 would then be joined together and split by Burly and Greedy. But a parcel consisting of s_1 and s_4 is only worth $440,000 to Greedy and half of that is only worth $220,000.

79. Arne gets the stereo and the table and pays $298.75; Brent patches the nail holes, repairs the window and gets $6.25 and the couch; Carlos gets with $251.25; Dale cleans the rugs and gets the desk and $41.25.

Item	**Arne**	**Brent**	**Carlos**	**Dale**	
Stereo	**300**	250	200	280	
Couch	200	**350**	300	100	
Table	**250**	200	240	80	
Desk	150	150	200	**220**	
Cleaning rugs	-80	-70	-100	**-60**	
Patching holes	-60	**-30**	-60	-40	
Repairing window	-60	**-50**	-80	-80	
Total Bids	700	800	700	500	
Fair Share	175	200	175	125	**Total**
Value of items received	550	270	0	160	**Surplus**
Prelim. cash settlement	-375	-70	175	-35	**305**
Share of surplus	76.25	76.25	76.25	76.25	
Final cash settlement	-298.75	6.25	251.25	41.25	

Chapter 4

WALKING

A. Standard Divisors and Standard Quotas

1. **(a)** Standard divisor $= \dfrac{3,310,000 + 2,670,000 + 1,330,000 + 690,000}{160} = 50,000$

(b) Apure: $\dfrac{3,310,000}{50,000} = 66.2$; Barinas: $\dfrac{2,670,000}{50,000} = 53.4$;

Carabobo: $\dfrac{1,330,000}{50,000} = 26.6$; Dolores: $\dfrac{690,000}{50,000} = 13.8$

3. **(a)** Standard divisor $= \dfrac{45,300 + 31,070 + 20,490 + 14,160 + 10,260 + 8,720}{130} = 1000$

(b) The "states" in any apportionment problem are the entities that will have "seats" assigned to them according to a share rule. In this case, the states are the 6 bus routes and the seats are the 130 buses. The standard divisor represents the average number of passengers per bus per day.

(c) A: $\dfrac{45,300}{1000} = 45.3$; B: $\dfrac{31,070}{1000} = 31.07$; C: $\dfrac{20,490}{1000} = 20.49$;
D: $\dfrac{14,160}{1000} = 14.16$; E: $\dfrac{10,260}{1000} = 10.26$; F: $\dfrac{8,720}{1000} = 8.72$

5. **(a)** Number of seats $= 40.50 + 29.70 + 23.65 + 14.60 + 10.55 = 119$

(b) Standard divisor $= \dfrac{23,800,000}{119} = 200,000$

(c) Population = standard quota × standard divisor, so
A: $40.5 \times 200,000 = 8,100,000$; B: $29.7 \times 200,000 = 5,940,000$
C: $23.65 \times 200,000 = 4,730,000$; D: $14.60 \times 200,000 = 2,920,000$
E: $10.55 \times 200,000 = 2,110,000$

7. With 7.43% of the U.S. population, Texas should receive 7.43% of the number of seats available in the House of Representatives. So, the standard quota for Texas is $0.0743 \times 435 = 32.3205$.

9. **(a)** Standard divisor $= \dfrac{100\%}{200} = 0.5\%$

(b)

State	A	B	C	D	E	F
Standard Quota	22.74	16.14	77.24	29.96	20.84	33.08

B. Hamilton's Method

11. Lower Quotas are: $A:66; B:53; C:26; D:13$, and the sum is 158. So we have 160–158 = 2 seats remaining to allocate. These are given to D and C, since they have the largest fractional parts of the standard quota. The final apportionment is: A: 66; B: 53; C: 27; D: 14.

13. Lower Quotas are: *A*: 45; *B*: 31; *C*: 20; *D*: 14; *E*: 10; *F*: 8, and the sum is 128. So we have 130–128 = 2 buses remaining to allocate. These are given to *F* and *C*, since they have the largest fractional parts of the standard quota. The final apportionment is: *A*: 45; *B*: 31; *C*: 21; *D*: 14; *E*: 10; *F*: 9.

15. Lower Quotas are: *A*: 40; *B*: 29; *C*: 23; *D*: 14; *E*: 10, and the sum is 116. So we have 119–116 = 3 seats remaining to allocate. These are given to *B*, *C* and *D*, since they have the largest fractional parts of the standard quota. The final apportionment is: *A*: 40; *B*: 30; *C*: 24; *D*: 15; *E*: 10.

17. Lower Quotas are: *A*: 22; *B*: 16; *C*: 77; *D*: 29; *E*: 20, *F*: 33, and the sum is 197. So we have 200–197 = 3 seats remaining to allocate. These are given to *A*, *D* and *E*, since they have the largest fractional parts of the standard quota. The final apportionment is: *A*: 23; *B*: 16; *C*: 77; *D*: 30; *E*: 21; *F*: 33.

19. (a)

Child	Bob	Peter	Ron
Standard quota	0.594	2.673	7.733
Lower quota	0	2	7

Note: standard divisor: $= \frac{54+243+703}{11} = 90.\overline{90}$

The sum of lower quotas is 9, so there are 2 remaining pieces of candy to allocate. These are given to Ron and Peter, since they have the largest fractional parts of the standard quota. The final apportionment is: Bob: 0; Peter: 3; Ron: 8.

(b)

Child	Bob	Peter	Ron
Study time	56	255	789
Standard quota	.56	2.55	7.89
Lower quota	0	2	7

Note: standard divisor $= \frac{56+255+789}{11} = 100$

The sum of the lower quotas is 9, so there are 2 remaining pieces of candy to allocate. These are given to Ron and Bob, since they have the largest fractional parts of the standard quota. The final apportionment is: Bob: 1; Peter: 2; Ron: 8.

(c) Yes. For studying an extra 2 minutes (an increase of 3.70%), Bob gets a piece of candy. However, Peter, who studies an extra 12 minutes (an increase of 4.94%), has to give up a piece. This is an example of the population paradox.

21. (a)

Child	Bob	Peter	Ron
Standard quota	.594	2.673	7.733
Lower quota	0	2	7

Note: standard divisor: $= \frac{54+243+703}{11} = 90.\overline{90}$

The sum of lower quotas is 9, so there are 2 remaining pieces of candy to allocate. These are given to Ron and Peter, since they have the largest fractional parts of the standard quota. The final apportionment is: Bob: 0; Peter: 3; Ron: 8.

(b)

Child	Bob	Peter	Ron	Jim
Study time	54	243	703	580
Standard quota	0.58	2.61	7.56	6.24
Lower quota	0	2	7	6

Note: standard divisor $= \dfrac{54+243+703+580}{17} \approx 92.94$

The sum of the lower quotas is 15, so there are 2 remaining pieces of candy to allocate. These are given to Peter and Bob, since they have the largest fractional parts of the standard quota. The final apportionment is: Bob: 1; Peter: 3; Ron: 7; Jim: 6.

(c) Ron loses a piece of candy to Bob when Jim enters the discussion and is given his fair share (6 pieces) of candy. This is an example of the new-states paradox.

C. Jefferson's Method

23. Any modified divisor between approximately 49,285.72 and 49,402.98 can be used for this problem. Using $D = 49{,}300$ we obtain:

State	*A*	*B*	*C*	*D*
Modified Quota	67.14	54.16	26.98	13.996
Modified Lower Quota	67	54	26	13

The final apportionment is the modified lower quota value in the table.

25. Any modified divisor between approximately 971 and 975.7 can be used for this problem. Using $D = 975$ we obtain:

Route	*A*	*B*	*C*	*D*	*E*	*F*
Modified Quota	46.46	31.87	21.02	14.52	10.52	8.94
Modified Lower Quota	46	31	21	14	10	8

The final apportionment is the modified lower quota value in the table.

27. Any modified divisor between approximately 194,666.67 and 197,083.33 can be used for this problem. Using $D = 195{,}000$ we obtain:

State	*A*	*B*	*C*	*D*	*E*
Modified Quota	41.54	30.46	24.26	14.97	10.82
Modified Lower Quota	41	30	24	14	10

The final apportionment is the modified lower quota value in the table.

29. Any modified divisor between approximately 0.4944% and 0.4951% can be used for this problem. Using $D = 0.495\%$ we obtain:

State	*A*	*B*	*C*	*D*	*E*	*F*
Modified Quota	22.9697	16.3030	78.0202	30.2626	21.0505	33.4141
Modified Lower Quota	22	16	78	30	21	33

The final apportionment is the modified lower quota value in the table.

31. From Exercise 7, the standard quota of Texas was 32.32 for the 2000 Census. However, Jefferson's method only allows for violations of the upper quota. This means that Texas could not receive fewer than 32 seats under Jefferson's method.

D. Adams' Method

33. Any modified divisor between approximately 50,377.4 and 50,923 can be used for this problem. Using D = 50,500 we obtain:

State	*A*	*B*	*C*	*D*
Modified Quota	65.54	52.87	26.34	13.66
Modified Upper Quota	66	53	27	14

The final apportionment is the modified upper quota value in the table.

35. Any modified divisor between approximately 1024.5 and 1026 can be used for this problem. Using D = 1025 we obtain:

Route	*A*	*B*	*C*	*D*	*E*	*F*
Modified Quota	44.20	30.31	19.99	13.81	10.01	8.51
Modified Upper Quota	45	31	20	14	11	9

The final apportionment is the modified upper quota value in the table.

37. Any modified divisor between approximately 204,828 and 205,652 can be used for this problem. Using D = 205,000 we obtain:

State	*A*	*B*	*C*	*D*	*E*
Modified Quota	39.51	28.98	23.07	14.24	10.29
Modified Upper Quota	40	29	24	15	11

The final apportionment is the modified upper quota value in the table.

39. Any modified divisor between approximately 0.5044% and 0.5081% can be used for this problem. Using D = 0.505% we obtain:

State	*A*	*B*	*C*	*D*	*E*	*F*
Modified Quota	22.5149	15.9802	76.4752	29.6634	20.6337	32.7525
Modified Upper Quota	23	16	77	30	21	33

The final apportionment is the modified upper quota value in the table.

41. The difference between 50 (the number of seats that California would receive under Adams' method) and 52.45 (California's standard quota) is greater than 1. This fact illustrates that Adams' method violates the quota rule.

E. Webster's Method

43. Any modified divisor between approximately 49,907 and 50,188 can be used for this problem. Using D = 50,000 we obtain:

State	*A*	*B*	*C*	*D*
Modified Quota	66.2	53.4	26.6	13.8
Rounded Quota	66	53	27	14

The final apportionment is the rounded quota value in the table.

45. Any modified divisor between approximately 995.61 and 999.51 can be used for this problem. Using $D = 996$ we obtain:

Route	*A*	*B*	*C*	*D*	*E*	*F*
Modified Quota	45.48	31.19	20.57	14.22	10.30	8.76
Rounded Quota	45	31	21	14	10	9

The final apportionment is the rounded quota value in the table.

47. Any modified divisor between approximately 200,953 and 201,276 can be used for this problem. Using $D =$ 201,000 we obtain:

State	*A*	*B*	*C*	*D*	*E*
Modified Quota	40.30	29.55	23.53	14.53	10.498
Rounded Quota	40	30	24	15	10

The final apportionment is the rounded quota value in the table.

49. Any modified divisor between approximately 0.499% and 0.504% can be used for this problem. Using $D =$ 0.5% we obtain:

State	*A*	*B*	*C*	*D*	*E*	*F*
Modified Quota	22.74	16.14	77.24	29.96	20.84	33.08
Rounded Quota	23	16	77	30	21	33

The final apportionment is the rounded quota value in the table.

JOGGING

51. (a) $q_1 + q_2 + \ldots + q_N$ represents the total number of seats available.

(b) $\dfrac{P_1 + P_2 + \ldots + P_N}{q_1 + q_2 + \ldots + q_N}$ represents the total population divided by the total number of seats available which also happens to be the standard divisor.

(c) $\dfrac{P_N}{P_1 + P_2 + \ldots + P_N}$ represents the percentage of the total population in the *N*th state.

53. If the modified divisor used in Jefferson's method is *D*, then state *A* has modified quota $\dfrac{P_A}{D} \geq 3.0$. Also, state *B* has modified quota $\dfrac{P_B}{D} < 1.0$. So, $P_A \geq 3D > 3P_B$. That is, more than $\dfrac{3}{4}$ of the total population lives in state *A*.

55. (a) Since the two fractional parts add up to 1, the surplus to be allocated using Hamilton's method is 1, and it will go to the state with larger fractional part, that is, the state with fractional part more than 0.5. This is the same result that is obtained by just rounding off in the conventional way, which in this case happens to be the result given by Webster's method. (Anytime that rounding off the quotas the conventional way produces integers that add up to M, Webster's method reduces to rounding off the standard quotas in the conventional way.)

(b) When there are only two states, Hamilton's and Webster's methods agree and Webster's method can never suffer from the Alabama or population paradox, hence neither will Hamilton's method.

(c) When there are only two states, Webster's and Hamilton's methods agree and Hamilton's method can never violate the quota rule, hence neither will Webster's method.

57. (a) The standard divisor is given by $\frac{1,262,505}{7.671} \approx 164,581.54$. This means each representative represents (roughly) 164,582 people. Since the number of seats is $M = 300$, the U.S. population can be estimated as $300 \times 164,581.54 \approx 49,374,462$.

(b) Since the standard quota for Texas is defined as the population of Texas divided by the standard divisor, it follows that the population of Texas is the product of its standard quota (9.672) and the standard divisor (164,581.54). That is, the population of Texas is $9.672 \times 164,581.54 \approx 1,591,833$.

59. (a) Any modified divisor between approximately 105.3 and 107.1 can be used for this problem. Using $D = 106$ we obtain:

State	*A*	*B*	*C*	*D*
Modified Quota	4.72	9.43	14.15	18.87
Modified Upper Quota	5	10	15	19

The final apportionment is the modified upper quota value in the table.

(b) For $D = 100$, the modified quotas are A: 5, B: 10, C: 15, D: 20 which sum to 50 seats. For $D < 100$, each of the modified quotas will increase, so rounding downward will give at least A: 5, B: 10, C: 15, D: 20 for a total of at least 50 seats. But for $D > 100$, *each* of the modified quotas will decrease, and so rounding downward will give at most A: 4, B: 9, C: 14, D: 19 for a total of 46 at most.

(c) From part (b), we see that there is no divisor such that the sum of the modified upper quotas will be 49.

61. Of those states for which the standard quota is not an integer, there are K that receive their lower quota. The remaining states (those with the largest fractional parts) receive their upper quota of seats. The states receiving their upper and lower quotas are the same under Hamilton's method or this alternate version of Hamilton's method.

63. (a) Take for example $q_1 = 3.9$ and $q_2 = 10.1$ (with $m = 14$). Under both Hamilton's method and Lowndes' method, A gets 4 seats and B gets 10 seats.

(b) Take for example $q_1 = 3.4$ and $q_2 = 10.6$ (with $m = 14$). Under Hamilton's method, A gets 3 seats and B gets 11 seats. Under Lowndes' method, A gets 4 seats and B gets 10 seats.

(c) Assume that $f_1 > f_2$, so under Hamilton's method the surplus seat goes to A. Under Lowndes' method, the surplus seat would go to B if $\frac{f_2}{q_2 - f_2} > \frac{f_1}{q_1 - f_1}$ which can be simplified to $f_2(q_1 - f_1) > f_1(q_2 - f_2)$ or $f_2 q_1 - f_2 f_1 > f_1 q_2 - f_1 f_2$ which gives $f_2 q_1 > f_1 q_2$. This can be rewritten as $\frac{q_1}{q_2} > \frac{f_1}{f_2}$ since all values are > 0.

Mini-Excursion 1

WALKING

A. The Geometric Mean

1. (a) $G = \sqrt{10 \times 100} = \sqrt{10,000} = 100$

(b) $G = \sqrt{20 \times 2000} = \sqrt{40,000} = 200$

(c) $G = \sqrt{\frac{1}{20} \times \frac{1}{2000}} = \sqrt{\frac{1}{40,000}} = \frac{1}{\sqrt{40,000}} = \frac{1}{200}$

3. $G = \sqrt{2 \cdot 3^4 \cdot 7^3 \cdot 11 \cdot 2^7 \cdot 7^5 \cdot 11} = \sqrt{2^8 \cdot 3^4 \cdot 7^8 \cdot 11^2} = 2^4 \cdot 3^2 \cdot 7^4 \cdot 11$

5. Approximately 18.6%
To compute the average annual increase in home prices, we first find the geometric mean of 1.117 and 1.259. This is given by $G = \sqrt{1.117 \times 1.259} = \sqrt{1.406303} \approx 1.186$. So, the average annual increase in home prices is approximately 18.6%.

7.

Consecutive integers	Geometric mean (G)	Arithmetic mean (A)	Difference (A-G)
15,16	15.492	15.5	0.008
16,17	16.492	16.5	0.008
17,18	17.493	17.5	0.007
18,19	18.493	18.5	0.007
19,20	19.494	19.5	0.006
29,30	29.496	29.5	0.004
39,40	39.497	39.5	0.003
49,50	49.497	49.5	0.003

9. (a) Since $G = \sqrt{a \cdot b}$, the geometric mean of $k \cdot a$ and $k \cdot b$ is $\sqrt{ka \cdot kb} = k\sqrt{a \cdot b} = kG$.

(b) Since $G = \sqrt{a \cdot b}$, the geometric mean of $\frac{k}{a}$ and $\frac{k}{b}$ is $\sqrt{\frac{k}{a} \cdot \frac{k}{b}} = k\frac{1}{\sqrt{a \cdot b}} = \frac{k}{G}$.

11. (a) Since $b > a$ and

$$\begin{aligned} |AC|^2 - |AB|^2 &= \left(\frac{b+a}{2}\right)^2 - \left(\frac{b-a}{2}\right)^2 \\ &= \frac{b^2 + 2ab + a^2 - b^2 + 2ab - a^2}{4} \\ &= ab \end{aligned}$$

the length of BC is given by $\sqrt{ab}$.

(b) Since the length of the legs of a right triangle are always strictly less than the length of the hypotenuse, it follows that $\sqrt{ab} < \frac{a+b}{2}$.

B. The Huntington-Hill Method

13. In this example the Huntington-Hill method produces the same apportionment as Webster's method and in fact, the standard divisor can be used as an appropriate divisor.

State	*A*	*B*	*C*	*D*	*E*	Total
Std. Quota	25.26	18.32	2.58	37.16	40.68	124
Geo. Mean (cutoff)	$\sqrt{25\times 26}\approx$ 25.4951	$\sqrt{18\times 19}\approx$ 18.4932	$\sqrt{2\times 3}\approx$ 2.4495	$\sqrt{37\times 38}\approx$ 37.4967	$\sqrt{40\times 41}\approx$ 40.4969	
Apportionment	25	18	3	37	41	124

15. **(a)** Under Webster's method the standard divisor $D = 10{,}000$ works!

State	*A*	*B*	*C*	*D*	*E*	*F*	Total
Population	344,970	408,700	219,200	587,210	154,920	285,000	2,000,000
Standard Quota	34.497	40.87	21.92	58.721	15.492	28.5	
Apportionment	34	41	22	59	15	29	200

(b) Under the Huntington-Hill method the standard divisor does not work! However, a modified divisor of $D = 10{,}001$ does work.

State	*A*	*B*	*C*	*D*	*E*	*F*	Total
Population	344,970	408,700	219,200	587,210	154,920	285,000	2,000,00
Mod. Quota ($D = 10{,}001$)	34.494	40.866	21.918	58.715	15.490	28.497	
Geo. Mean	34.496	40.497	21.494	58.498	15.492	28.496	
Apportionment	34	41	22	59	15	29	200

(c) The apportionments came out the same for both methods.

17. **(a)** Under the Huntington-Hill method the standard divisor does not work but a modified divisor of $D = 990$ works!

State	*Aleta*	*Bonita*	*Corona*	*Doritos*	Total
Population	86,915	4,325	5,400	3,360	100,000
Modified Quota ($D = 990$)	87.793	4.369	5.455	3.394	
Geo. Mean	87.4986	4.4721	5.4772	3.4641	
Apportionment	88	4	5	3	100

(b) The apportionment using the Huntington-Hill method violates the quota rule (an upper quota violation). Aleta has a standard quota of 86.915 and yet receives 88 representatives in the final apportionment.

C. Miscellaneous

19. If $b < c$, then $\sqrt{b} < \sqrt{c}$. It follows that $\sqrt{c} - \sqrt{a} > \sqrt{b} - \sqrt{a}$. Squaring both sides yields $\left(\sqrt{c} - \sqrt{a}\right)^2 > \left(\sqrt{b} - \sqrt{a}\right)^2$. Multiplying this out gives $c - 2\sqrt{ac} + a > b - 2\sqrt{ab} + a$. Dividing both sides by 2 and rearranging terms produces the inequality we are seeking: $\frac{a+c}{2} - \sqrt{ac} > \frac{a+b}{2} - \sqrt{ab}$.

21. (a)

a	*b*	*Q*	*a*	*b*	*Q*
1	2	1.581	9	10	9.513
2	3	2.550	19	20	19.506
3	4	3.536	29	30	29.504
4	5	4.528	99	100	99.501
5	6	5.523	10	20	15.811
6	7	6.519	10	90	64.031
7	8	7.517	10	190	134.536
8	9	8.515	10	990	700.071

(b) Since $a < b$, and both a and b are positive, it follows that $a^2 = \frac{a^2 + a^2}{2} < \frac{a^2 + b^2}{2} < \frac{b^2 + b^2}{2} = b^2$. Taking square roots and using the fact that a and b are positive, $a < \sqrt{\frac{a^2 + b^2}{2}} < b$.

(c) $Q = \sqrt{\frac{a^2 + b^2}{2}} = \sqrt{\left(\frac{b-a}{2}\right)^2 + \left(\frac{b+a}{2}\right)^2} \geq \sqrt{\left(\frac{b+a}{2}\right)^2} = \frac{b+a}{2}$

Chapter 5

WALKING

A. Graphs: Basic Concepts

1. (a) V = {*A, B, C, X, Y, Z*}
The basic elements of a graph are dots and lines (vertices and edges in mathematical language). The vertex set is a listing of the labels given to each dot (vertex).

(b) E = {*AX, AY, AZ, BB, BC, BX, CX, XY*}
Remember, each edge can be described by listing the pair of vertices that are connected by that edge.

(c) deg(*A*) = 3, deg(*B*) = 4, deg(*C*) = 2, deg(*X*) = 4, deg(*Y*) = 2, deg(*Z*) = 1
The degree of a vertex is the number of edges meeting at that vertex.

3. (a) V = {*A, B, C, D, X, Y, Z*}

(b) E = {*AX, AX, AY, BX, BY, DZ, XY*}
Note: *AX* is listed twice.

(c) deg(*A*) = 3, deg(*B*) = 2, deg(*C*) = 0, deg(*D*) = 1, deg(*X*) = 4, deg(*Y*) = 3, deg(*Z*) = 1

5. One possible picture of the graph:

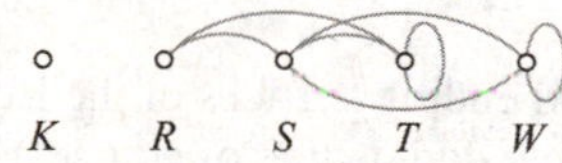

A second possible picture:

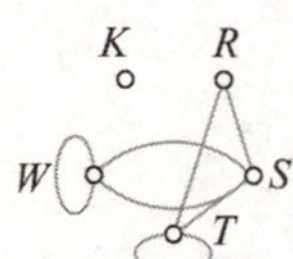

7. (a) *A, B, D, E*
Remember that two vertices are adjacent in a graph if they are joined by an edge.

(b) *AD, BC, DD, DE*
Two edges are adjacent if they share a common vertex.

(c) 5

(d) 12; one way to easily find this is to multiply the number of edges by two (since each edge contributes two degrees to the total).

9. (a)

(b)

(c)

11. (a) *C, B, A, H, F*
Other answers such as *C, B, A, H, G, F* are also possible.

(b) *C, B, D, A, H, F*
Again, more than one answer is possible.

(c) *C, B, A, H, F*

(d) *C, D, B, A, H, G, G, F*

(e) 4 (It can be helpful to list them from shortest to longest as *C, B, A*; *C, D, A*; *C, B, D, A*; *C, D, B, A*)

(f) 3 (*H, F*; *H, G, F*; *H, G, G, F*)

(g) 12 (Any one of the 4 paths in (e) followed by edge *AH* followed by any one of the 3 paths in (f).)

13. (a) *G,G.* A circuit of length 1 is a loop.

(b) There are none. Such a circuit would consist of two vertices and two (different) edges connecting the vertices.

(c) *A,B,D,A; B,C,D,B; F,G,H,F*

(d) *A,B,C,D,A; F,G,G,H,F*

(e) 6; The longest circuit has length 4. So using the results from (a)-(d) there are a total of 6 circuits in the graph.

15. (a) *AH, EF*; If either of these edges were removed from the graph, the graph would be disconnected.

(b) There are none. This graph is just one circuit connecting five vertices. Circuits do not have bridges.

(c) *AB, BC, BE, CD*; That is, every edge is a bridge.

B. Graph Models

17. Such a graph will have five vertices (representing North Kingsburg, South Kingsburg, and islands *A*, *B*, and *C*) and seven edges (representing the seven bridges).

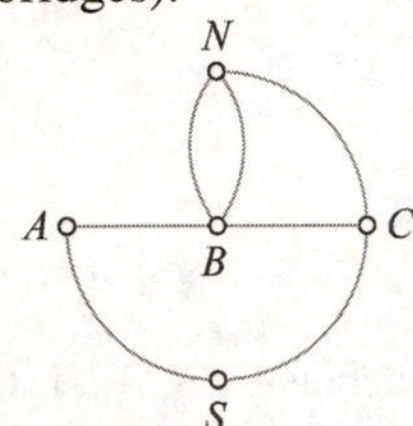

19.

21. Let the vertices represent the teams *A*, *B*, *C*, *D*, *E*, and *F*. The edges will correspond to the tournament pairings. Putting the vertices at the corners of a regular hexagon will make drawing and interpreting the graph a bit easier.

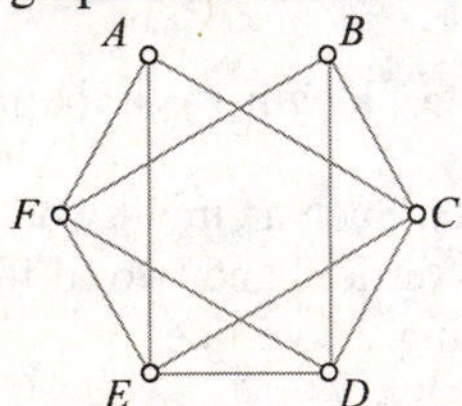

C. Euler's Theorems

23. (a) (i); The graph has an Euler circuit because all vertices have even degree.

(b) (iii); The graph has neither an Euler circuit nor an Euler path because there are four vertices of odd degree.

(c) (iv); See exercises 9(a) where an Euler circuit exists and 9(b) where an Euler circuit does not exist.

25. (a) (iii); The graph has neither an Euler circuit nor an Euler path because there are more than two (exactly 4 in fact) vertices of odd degree.

(b) (i); The graph has an Euler circuit because all vertices have even degree.

(c) (iii); The graph is disconnected. Disconnected graphs have neither Euler circuits nor Euler paths.

27. (a) (ii); Exactly two vertices are odd.

(b) (i); All the vertices are even.

(c) (iii); More than two vertices (6 in this case) are odd.

D. Finding Euler Circuits and Euler Paths

29.

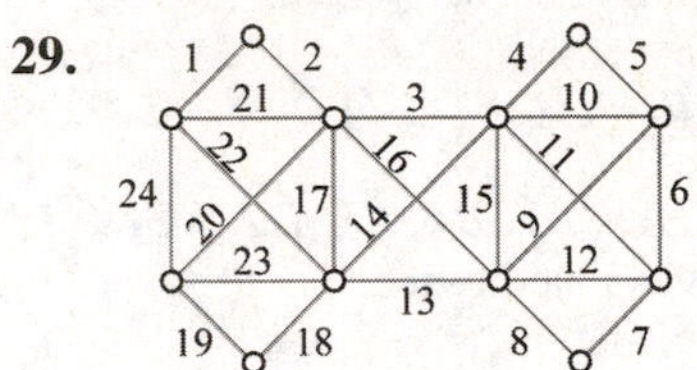

31.

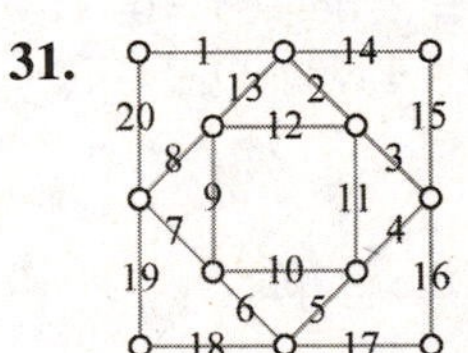

33. The starting and ending vertices of the Euler path (both having odd degree of course!) are shown in black. Naturally, other answers are possible.

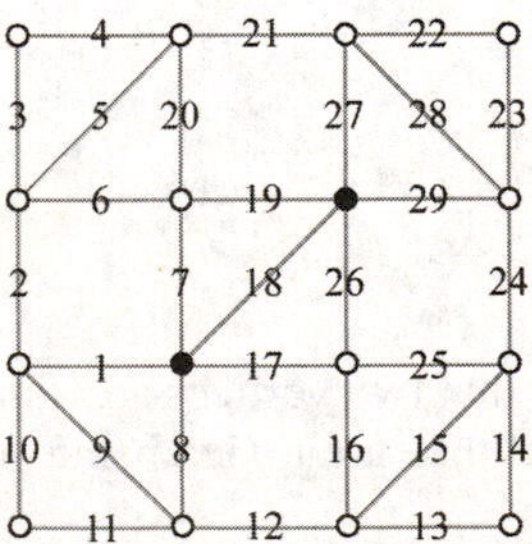

35. Again, there is more than one answer.

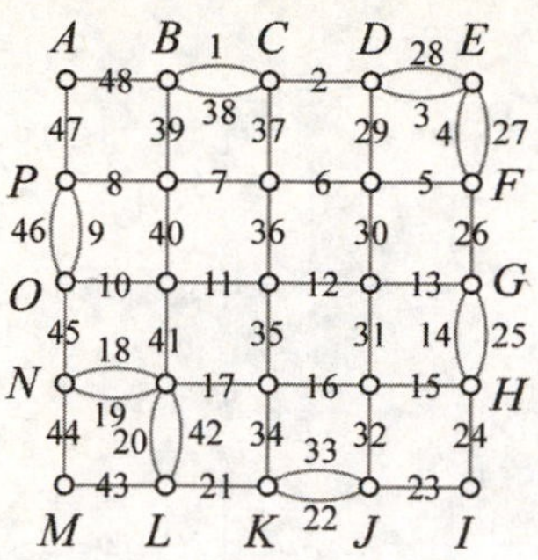

F. Eulerizations and Semi-eulerizations

37. Adding as few duplicate edges as possible in eliminating the odd vertices gives an optimal eulerization.

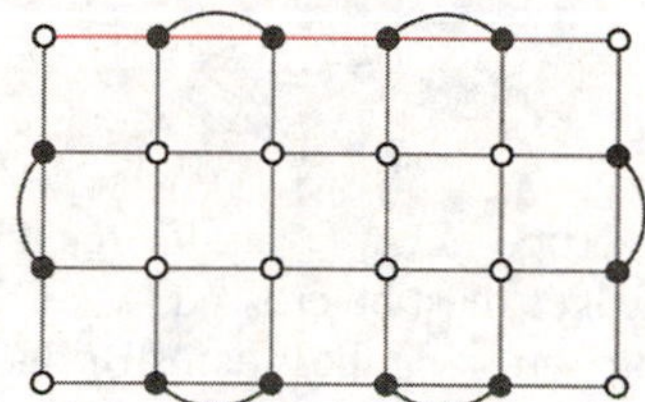

39. Since this graph has vertical symmetry, a mirror image of this answer is also an optimal eulerization.

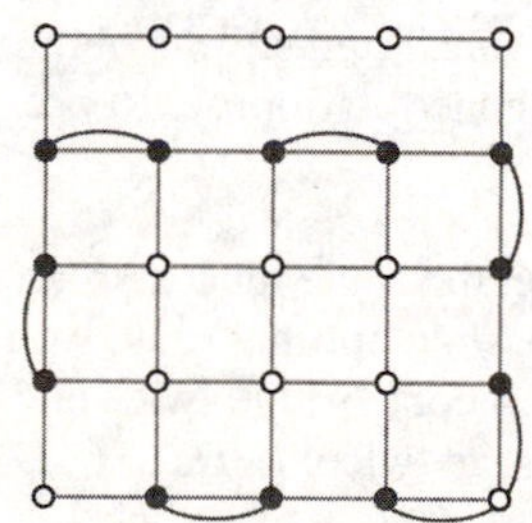

41. In an optimal semi-eulerization two vertices will remain odd. Again, since this graph has vertical symmetry, a mirror image of this answer is also an optimal semi-eulerization.

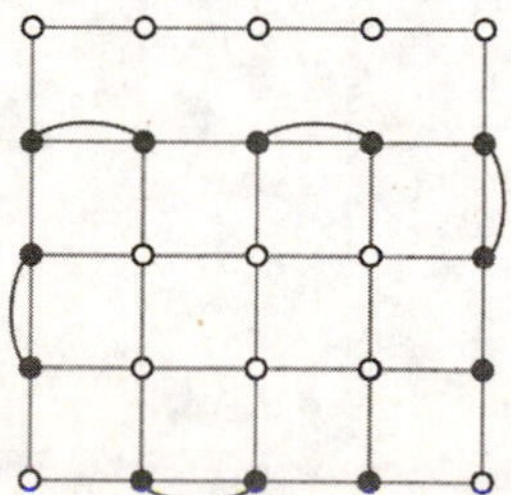

G. Miscellaneous

43. **(a)**

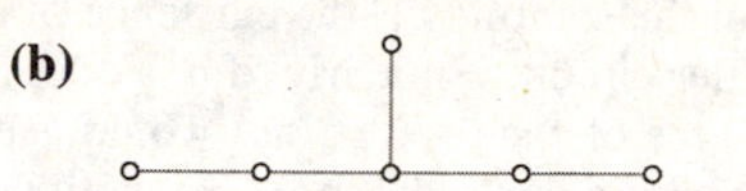

(b)

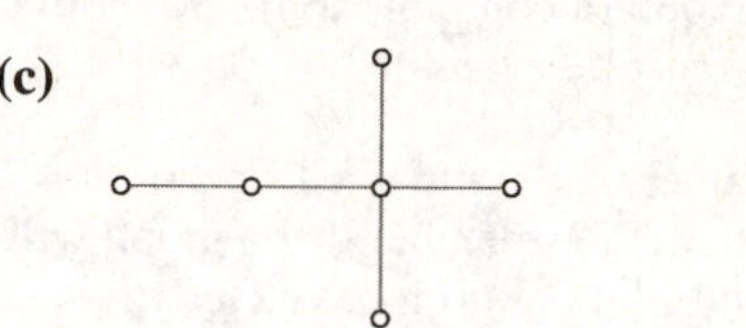

(c)

(d)

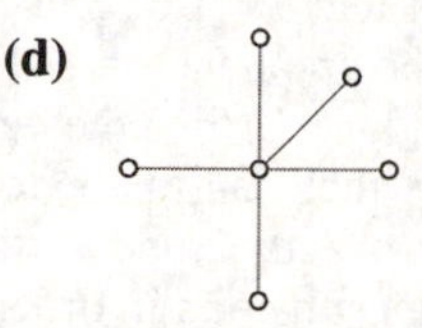

45.

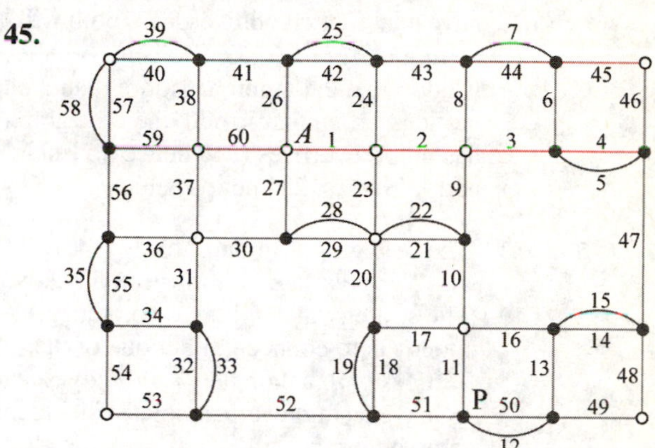

47. You would need to lift your pencil 4 times. There are 10 vertices of odd degree. Two can be used as the starting and ending vertices. The remaining eight odd degree vertices can be paired so that each pair forces one lifting of the pencil.

JOGGING

49. **(a)** An edge XY contributes 2 to the sum of the degrees of all vertices (1 to the degree of X, and 1 to the degree of Y).

(b) If there were an odd number of odd vertices, the sum of the degrees of all the vertices would be odd. That would contradict the result found in (a).

51. None. If a vertex had degree 1, then the single edge incident to that vertex would be a bridge.

53. **(a)** Both m and n must be even. Why? If the graph has an Euler circuit, then the degree of each degree must be even. Suppose that m is odd (i.e. A has an odd number of vertices). Then the degree of each vertex in B would be an odd number (namely m). But this contradicts the existence of an Euler circuit. Similarly, if n is odd, the degree of each vertex in A would need to be odd (a contradiction). So, both m and n must be even.

(b) Either $m = 1$ and $n = 1$ or 2 or $m = 2$ and n is odd. Remember that a graph will have an Euler path if it has exactly two odd vertices. If $m = 1$ and $n = 1$ or 2, then it is easy to see that the graph has an Euler path. If $m = 1$ and $n > 2$, then the vertices in B are all odd (which can't happen since an Euler path has exactly 2 odd vertices). If $m = 2$, then n must be odd in order to have exactly two odd vertices (both will be in A). If $m > 2$, then so is n. If either is odd, then the graph has more than 2 odd vertices. If neither is odd, then the graph has no odd vertices (and hence no Euler path). So $m > 2$ is not possible.

55. (a) The best graphs to build for $N = 4$, $N = 5$, and $N = 6$ vertices are illustrated below. The strategy that will give you the most money is to connect all but one of the vertices with a complete set of edges and to leave a single vertex isolated (incident to no edges).

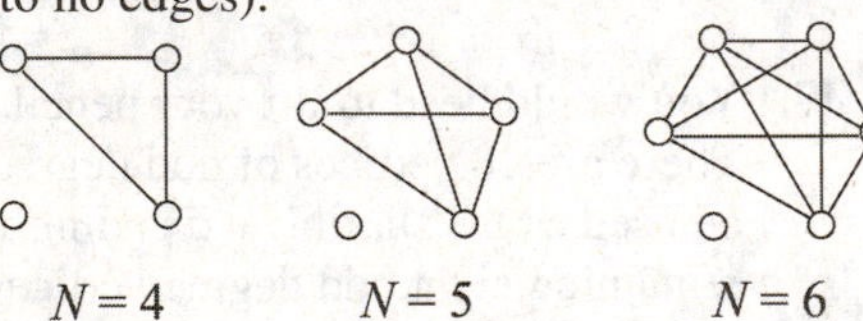

(b) Using the strategy found in part (a), the most money you can make in building a graph with N vertices is $\$[1+2+3+\ldots+(N-3)+(N-2)] = \$\frac{(N-2)(N-1)}{2}$. [For the first vertex, there are N-2 edges that can be built, for the second vertex, there are only N-3 edges that can be built, etc…]

57. (a) 12 (See edges that have been added in part (b).)

(b)

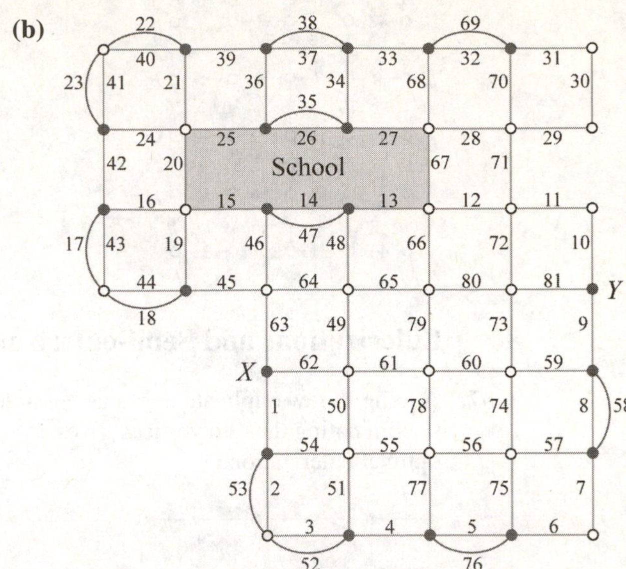

59. (a) Since vertices R, B, D, and L are all odd, two bridges must be crossed twice. For example, crossing the Adams Bridge twice and the Lincoln Bridge twice produces an optimal route: *D, L, R, L, C, A, R, B, R, B, D, C, L, D*.
Listing the bridges in the order they are crossed: Adams, Washington, Jefferson, Grant, Wilson, Truman, Lincoln, Lincoln, Hoover, Kennedy, Monroe, Roosevelt, Adams.

(b) In this case, the route must start and end at L. One possible optimal route would cross the Lincoln Bridge twice and the Adams Bridge twice: *L, D, L, C, D, B, R, B, R, L, R, A, C, L*.
Listing the bridges in the order they are crossed: Adams, Adams, Roosevelt, Monroe, Kennedy, Hoover, Lincoln, Lincoln, Washington, Jefferson, Truman, Wilson, Grant.

Chapter 6

WALKING

A. Hamilton Circuits and Hamilton Paths

1. (a) 1. A, B, D, C, E, F, G, A;
2. A, D, C, E, B, G, F, A;
3. A, D, B, E, C, F, G, A

(b) A, G, F, E, C, D, B

(c) D, A, G, B, C, E, F

3. (a) Starting at A, suppose that you travel to B. Then, there are two choices for which vertex to travel to next: C and E. If one starts out A,B,C, then the next vertex visited must be D (do you see why?). If one starts out A,B,E, then the next vertex visited must also be D. After D, there is only one way to return to A to form a Hamilton circuit. So the Hamilton circuits are A,B,C,D,E,F,A and A,B,E,D,C,F,A and their mirror images.

(b) The Hamilton circuits here will be the same as in (a) except that instead of traveling from F back to A, one must travel from F to G and then on to A. The Hamilton circuits are thus A,B,C,D,E,F,G,A and A,B,E,D,C,F,G,A and their mirror images.

(c) A,B,C,D,E,F,G,A; A,B,E,D,C,F,G,A; A,F,C,D,E,B,G,A; A,F,E,D,C,B,G,A and their mirror images.

The first two Hamilton circuits listed are exactly those found in (b). The second two Hamilton circuits are exactly the same as the first two with B and F reversing roles (since one can now access vertex G from B).

5. (a) A, F, B, C, G, D, E

(b) A, F, B, C, G, D, E, A

(c) A, F, B, E, D, G, C

(d) F, A, B, E, D, C, G

7. (a) Suppose a Hamilton circuit starts by moving to the right (to vertex B). At B there are two choices as to where to go next (C or E). Each of these choices completely determines the rest of the circuit.

Starting by moving to the right will force the circuit to end at F and then A. So the mirror image will represent the case of starting by moving left.

The two Hamilton circuits are then A, B, C, D, E, F, A and A, B, E, D, C, F, A and their mirror image circuits A, F, E, D, C, B, A and A, F, C, D, E, B, A.

(b) The circuits are the same as in (a). However, they are simply described in a different manner.
1. D, E, F, A, B, C, D
2. D, C, F, A, B, E, D
Mirror-image circuits:
3. D, C, B, A, F, E, D
4. D, E, B, A, F, C, D

9. The degree of every vertex in a graph with a Hamilton circuit must be at least 2 since the circuit must "pass through" every vertex. This graph has 2 vertices of degree 1.

Any Hamilton *path* passing through vertex A must contain edge AB. Any path passing through vertex E must contain edge BE. And, any path passing through vertex C must contain edge BC. Consequently, any path passing through vertices A, C, and E must contain at least three edges meeting at B and hence would pass through vertex B more than once. So, this graph does not have any Hamilton paths.

11. (a) $A, I, J, H, B, C, F, E, G, D$

(b) $G, D, E, F, C, B, A, I, J, H$

(c) If such a path were to start by heading left, it would not contain C, D, E, F, or G since it would need to pass through B again in order to do so. On the other hand, if a path were to start by heading right, it could not contain A, I H, or J.

(d) Any circuit would need to cross the bridge BC twice (in order to return to where it started). But then B and C would be included twice.

13. **(a)** 6

(b) *B, D, A, E, C, B*
Weight = 6 + 1 + 9 + 4 + 7 = 27

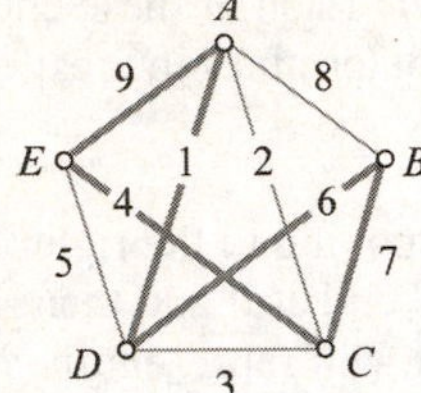

(c) The mirror image *B, C, E, A, D, B*
Weight = 7 + 4 + 9 + 1 + 6 = 27

15. **(a)** *A, D, F, E, B, C*
Weight = 2 + 7 + 5 + 4 + 11 = 29

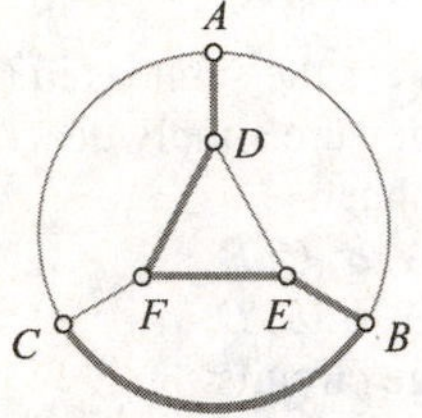

(b) *A, B, E, D, F, C*
Weight = 10 + 4 + 3 + 7 + 6 = 30

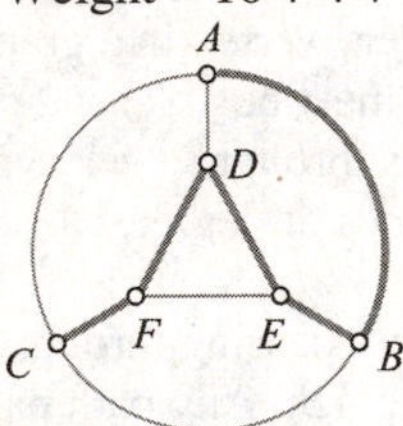

(c) There are only two Hamilton paths that start at *A* and end at *C*. The path *A, D, F, E, B, C* found in part (a) is the optimal such path. It has weight 29.

B. Factorials and Complete Graphs

17. **(a)** $10! = 10\times9\times8\times\ldots\times3\times2\times1$

$= 10\times9!$

So, $9! = \frac{10!}{10} = \frac{3,628,800}{10} = 362,880$.

(b) $11! = 11\times10\times9\times\ldots\times3\times2\times1 = 11\times10!$

So, $\frac{11!}{10!} = \frac{11\times10!}{10!} = 11$.

(c) $11! = 11\times10\times(9\times\ldots\times3\times2\times1)$

$= 11\times10\times9!$

So, $\frac{11!}{9!} = \frac{11\times10\times9!}{9!} = 11\times10 = 110$.

(d) $\frac{9!}{6!} = \frac{9\times8\times7\times6!}{6!} = 9\times8\times7 = 504$

(e) $\frac{101!}{99!} = \frac{101\times100\times99!}{99!} = 101\times100 = 10,100$

19. **(a)** $\frac{9!+11!}{10!} = \frac{9!}{10!} + \frac{11!}{10!}$

$= \frac{9!}{10\times9!} + \frac{11\times10!}{10!}$

$= \frac{1}{10} + 11$

$= 11.1$

(b) $\frac{101!+99!}{100!} = \frac{101!}{100!} + \frac{99!}{100!}$

$= \frac{101\times100!}{100!} + \frac{99!}{100\times99!}$

$= 101 + \frac{1}{100}$

$= 101.01$

21. **(a)** 2,432,902,008,176,640,000 $\approx 2.43\times10^{18}$

(b) 1,124,000,727,777,607,680,000
$\approx 1.12\times10^{21}$

(c) 20! = 2,432,902,008,176,640,000
$\approx 2.43\times10^{18}$

23. **(a)** $20! \text{ circuits} \times \frac{1 \text{ sec}}{1,000,000,000 \text{ circuits}}$

$\times \frac{1 \text{ min}}{60 \text{ sec}} \times \frac{1 \text{ hr}}{60 \text{ min}} \times \frac{1 \text{ day}}{24 \text{ hr}} \times \frac{1 \text{ year}}{365 \text{ days}}$

$\approx$ 77 years

(b) $21! \text{ circuits} \times \frac{1 \text{ sec}}{1,000,000,000 \text{ circuits}}$

$\times \frac{1 \text{ min}}{60 \text{ sec}} \times \frac{1 \text{ hr}}{60 \text{ min}} \times \frac{1 \text{ day}}{24 \text{ hr}} \times \frac{1 \text{ year}}{365 \text{ days}}$

$\approx$ 1620 years

25. **(a)** $\frac{20\times19}{2} = 190$

(b) K_{21} has 20 more edges than K_{20}.

$\frac{21\times20}{2} = 210$

(c) If one vertex is added to K_{50} (making for 51 vertices), the new complete graph K_{51} has 50 additional edges (one to each old vertex). In short, K_{51} has 50 more edges than K_{50}. That is, $y - x = 50$.

27. (a) K_N has $(N-1)!$ Hamilton circuits and 120 = 5!. So $N = 6$.

(b) K_N has $\frac{(N-1)N}{2}$ edges and 45 = $\frac{9\times 10}{2}$, so $N = 10$.

(c) $20{,}100 = \frac{200\times 201}{2}$, so $N = 201$.

C. Brute Force and Nearest Neighbor Algorithms

29. (a)

Tour (Mirror Image)	Weight
A, B, C, D, A (*A, D, C, B, A*)	48 + 32 + 18 + 22 = 120
A, B, D, C, A (*A, C, D, B, A*)	48 + 20 + 18 + 28 = 114
A, C, B, D, A (*A, D, B, C, A*)	28 + 32 + 20 + 22 = 102

An optimal tour would be *A, C, B, D, A* (or its mirror image) with a cost of 102.

(b) *A, D, C, B, A*
Weight = 22 + 18 + 32 + 48 = 120

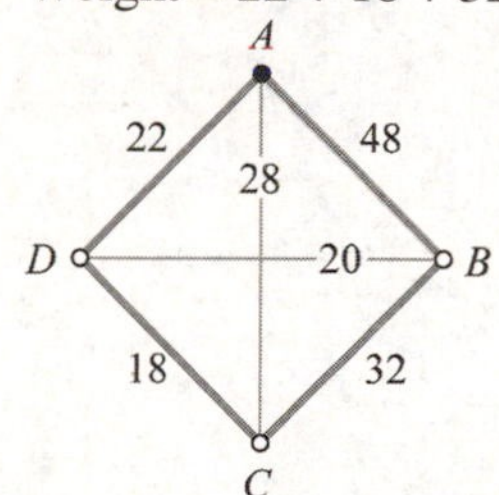

(c) *B, D, C, A, B*
Weight = 20 + 18 + 28 + 48 = 114

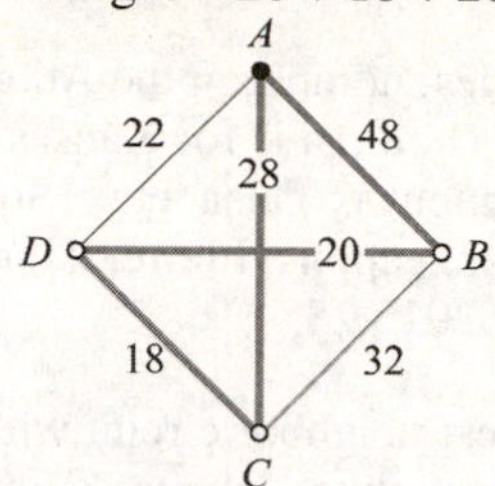

(d) *C, D, B, A, C*
Weight = 18 + 20 + 48 + 28 = 114

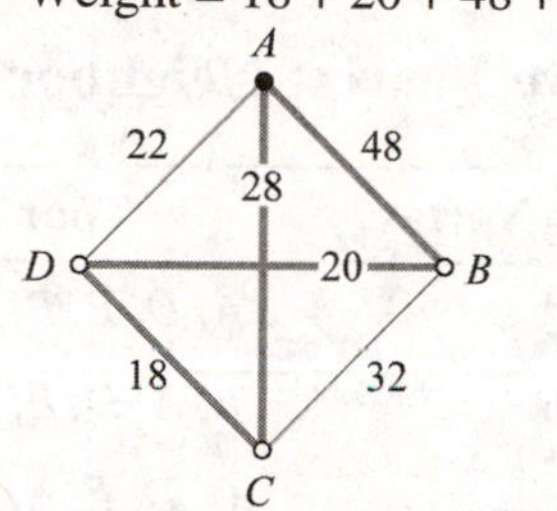

31. (a) *B, C, A, E, D, B*; Cost = \$121 + \$119 + \$133 + \$199 + \$150 = \$722

(b) *C, A, E, D, B, C*; Cost = \$119 + \$133 + \$199 + \$150 + \$121 = \$722

(c) *D, B, C, A, E, D*; Cost = \$150 + \$121 + \$119 + \$133 + \$199 = \$722

(d) *E, C, A, D, B, E*; Cost = \$120 + \$119 + \$152 + \$150 + \$200 = \$741

33. (a) *A, D, E, C, B, A*
Cost of bus tour = \$8 × (185 + 302 + 165 + 305 + 500) = \$11,656

(b) *A, D, B, C, E, A*
Cost of bus tour = \$8 × (185 + 360 + 305 + 165 + 205) = \$9,760

(c) There are only 6 possible tours that make *B* the first stop after *A*.

Tour	Cost
A, B, C, D, E, A	\$8 × (500 + 305 + 320 + 302 + 205) = \$13,056
A, B, C, E, D, A	\$8 × (500 + 305 + 165 + 302 + 185) = \$11,656
A, B, D, C, E, A	\$8 × (500 + 360 + 320 + 165 + 205) = \$12,400
A, B, D, E, C, A	\$8 × (500 + 360 + 302 + 165 + 200) = \$12,216
A, B, E, C, D, A	\$8 × (500 + 340 + 165 + 320 + 185) = \$12,080
A, B, E, D, C, A	\$8 × (500 + 340 + 302 + 320 + 200) = \$13,296

Optimal tour starting with A and then B: A, B, C, E, D, A
Cost of this bus tour = \$8× (500 + 305 + 165 + 302 + 185) = \$11,656

35. **(a)** The smallest number in the Atlanta column is for Columbus; the smallest number other than Atlanta in the Columbus column is for Kansas City; the smallest number in the Kansas City column other than Atlanta and Columbus is Tulsa, etc.. So the nearest-neighbor tour for Darren is Atlanta, Columbus, Kansas City, Tulsa, Minneapolis, Pierre, Atlanta. The cost of this trip = \$0.75× (533 + 656 + 248 + 695 + 394 + 1361) = \$2915.25.

(b) The nearest-neighbor circuit with Kansas City as the starting vertex is Kansas City, Tulsa, Minneapolis, Pierre, Columbus, Atlanta, Kansas City. Written with starting city Atlanta, the circuit is Atlanta, Kansas City, Tulsa, Minneapolis, Pierre, Columbus, Atlanta. The cost of this trip = \$0.75× (798 + 248 + 695 + 394 + 1071 + 533) = \$2804.25

D. Repetitive Nearest-Neighbor Algorithm

37.

Starting Vertex	Tour	Weight
A	A, D, E, B, C, A	3.1 + 2.2 + 3.8 + 3.6 + 3.3 = 16.0
B	B, A, D, E, C, B	3.2 + 3.1 + 2.2 + 4.1 + 3.6 = 16.2
C	C, D, E, A, B, C	2.4 + 2.2 + 3.4 + 3.2 + 3.6 = 14.8
D	D, E, A, B, C, D	2.2 + 3.4 + 3.2 + 3.6 + 2.4 = 14.8
E	E, D, C, A, B, E	2.2 + 2.4 + 3.3 + 3.2 + 3.8 = 14.9

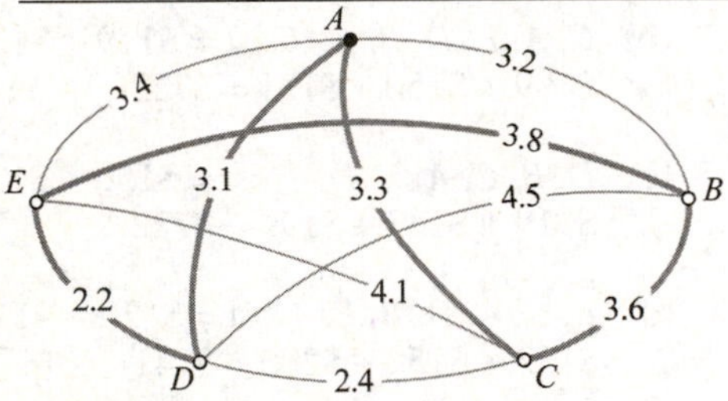

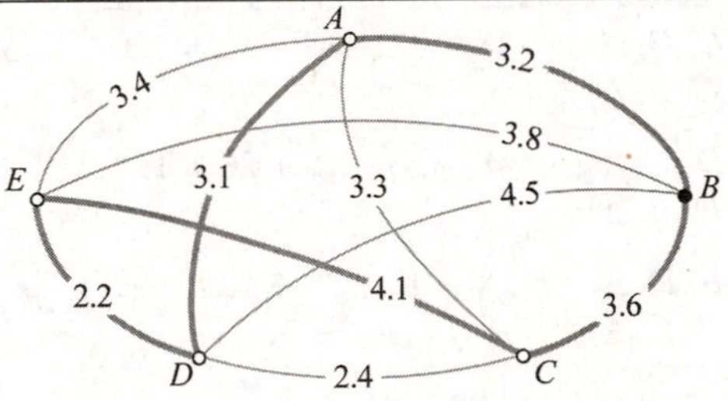

Weight: 16.0 16.2

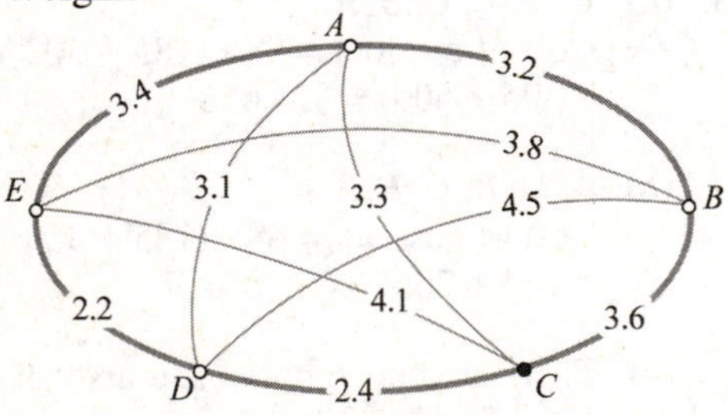

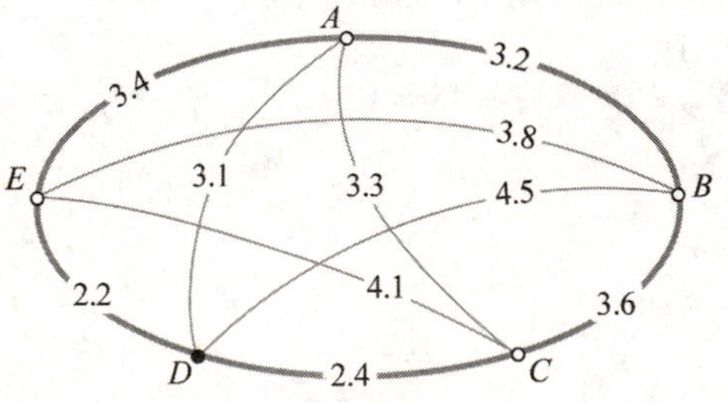

Weight: 14.8 14.8

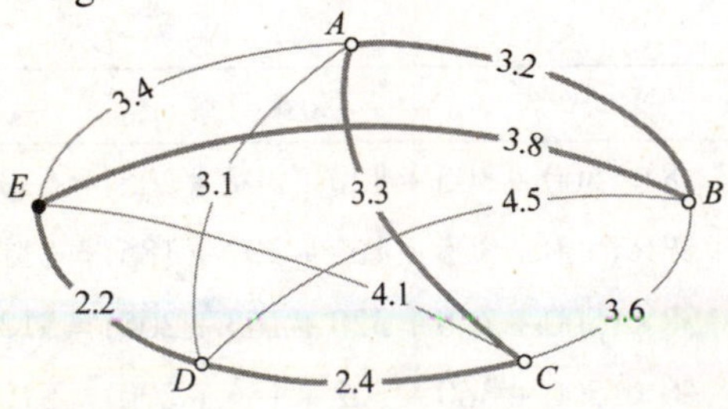

Weight: 14.9

Starting with vertex B, the shortest circuit is B, C, D, E, A, B with weight 14.8.

39.

Starting Vertex	Tour	Length of Tour (Miles)
A	*A, D, E, C, B, A*	185 + 302 + 165 + 305 + 500 = 1457
B	*B, C, E, A, D, B*	305 + 165 + 205 + 185 + 360 = 1220
C	*C, E, A, D, B, C*	165 + 205 + 185 + 360 + 305 = 1220
D	*D, A, C, E, B, D*	185 + 200 + 165 + 340 + 360 = 1250
E	*E, C, A, D, B, E*	165 + 200 + 185 + 360 + 340 = 1250

Weight: 1457 1220 1220 1250 1250

The shortest tour (written as starting and ending at *A*) is *A, D, B, C, E, A* with length 1220. The cost of this bus tour is $8× 1220 = $9,760.

41.

Starting Vertex	Tour	Length of Tour (Miles)
A	*A, C, K, T, M, P, A*	533 + 656 + 248 + 695 + 394 + 1361 = 3887
C	*C, A, T, K, M, P, C*	533 + 772 + 248 + 447 + 394 + 1071 = 3465
K	*K, T, M, P, C, A, K*	248 + 695 + 394 + 1071 + 533 + 798 = 3739
M	*M, P, K, T, A, C, M*	394 + 592 + 248 + 772 + 533 + 713 = 3252
P	*P, M, K, T, A, C, P*	394 + 447 + 248 + 772 + 533 + 1071 = 3465
T	*T, K, M, P, C, A, T*	248 + 447 + 394 + 1071 + 533 + 772 = 3465

The shortest circuit is *M, P, K, T, A, C, M*. Written starting from Atlanta, this is Atlanta, Columbus, Minneapolis, Pierre, Kansas City, Tulsa, Atlanta and has a length of 3252 miles. The total cost of this tour is $0.75× 3252 = $2439.

E. Cheapest-Link Algorithm

43. The cheapest-link tour is created by adding edges *DE, CD, AB, AC* and *BE* in that order. Edges *DE* and *CD* are added to the tour first since these are the cheapest two edges in the graph. Edge *AD* is the next cheapest edge. However, it is not added to the tour since then three edges would meet at vertex *D*. So, *AB* is the third edge added to the tour.

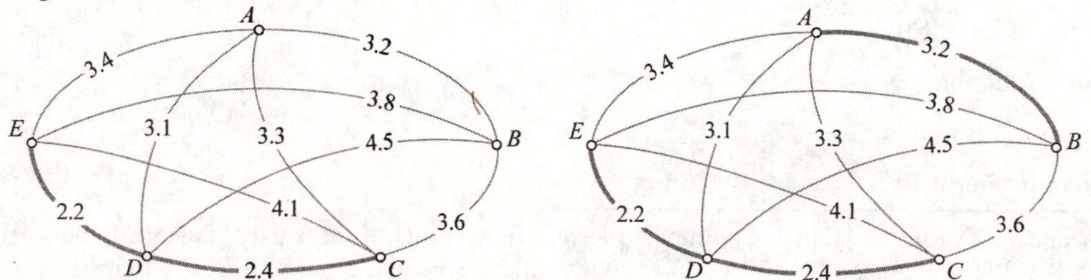

Edge *AC* is next in line (i.e. next smallest weight) to be added and doing so violates no rules in the cheapest-link algorithm. To complete the tour at this point we have no choice but to add edge *BE* (having weight 3.8).

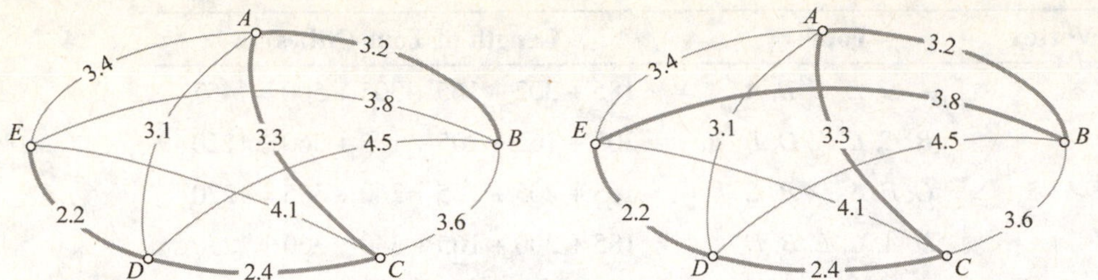

The shortest tour found using this algorithm is *B, E, D, C, A, B*. The weight of this tour is 2.2 + 2.4 + 3.2 + 3.3 + 3.8 = 14.9.

45. The first three steps in the cheapest-link algorithm are illustrated in the following figures.

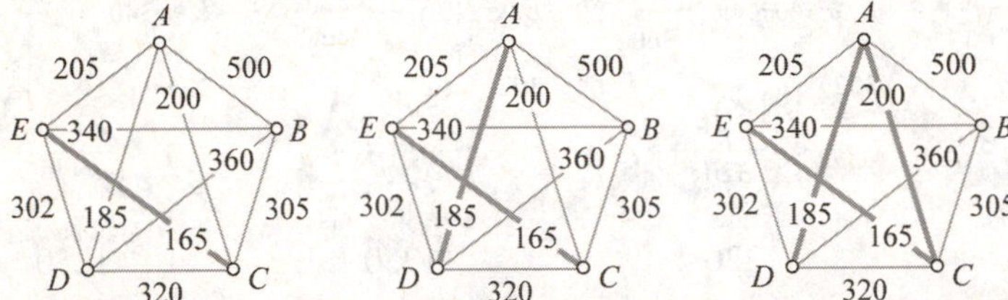

After the third step, the next cheapest edge is *AE* (weight 205). But that edge would make three edges come together at vertex *A* so we skip it and move to the next cheapest edge which is *DE* (weight 302). Since this edge makes a circuit, we skip this edge and try the next cheapest edge. The next cheapest edge is *BC* (weight 305) but that makes three edges come together at vertex *C* so we skip this edge and choose the next cheapest edge which is *CD* (weight 320). That edge makes a circuit and also makes three edges come together at vertex *C*. So, we skip it and choose the next cheapest edge which is *BE* (weight 340). After *BE* is added, only one edge, *BD*, will complete the tour.

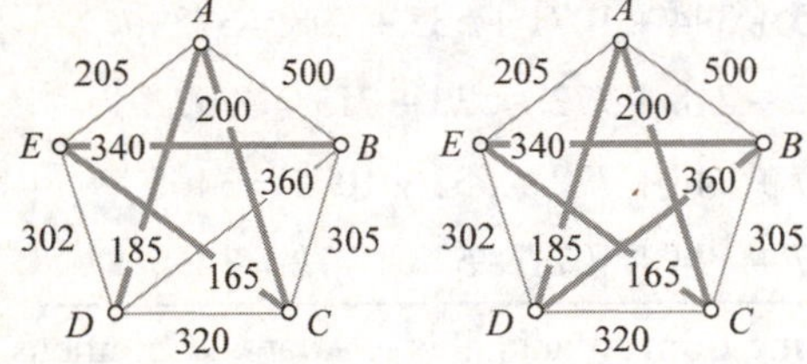

The shortest tour found using the cheapest-link algorithm is *A, D, B, E, C, A*. The cost of this bus tour is $8× (165 + 185 + 200 + 340 + 360) = $10,000.

47.

Link Added to Tour	Cost of Link (Miles)
Kansas City - Tulsa	248
Pierre - Minneapolis	394
Minneapolis - Kansas City	447
Atlanta - Columbus	533
Atlanta - Tulsa	772
Columbus - Pierre	1071

Darren's cheapest-link tour is Atlanta, Columbus, Pierre, Minneapolis, Kansas City, Tulsa, Atlanta. His total mileage is 248 + 394 + 447 + 533 + 772 + 1071 = 3465 miles. His total cost is $0.75 × 3465 = $2598.75.

F. Miscellaneous

49. The relative error of the nearest-neighbor tour is $\varepsilon = (W - Opt)/Opt$ where *W* is the weight of the nearest-neighbor tour and *Opt* is the weight of the optimal tour. In this case, we see that this error is $\varepsilon = (W - Opt)/Opt = (\$2153 - \$1914)/\$1914 = 12.5\%$

51. We solve $(\$1614 - Opt)/Opt = 7.6\%$ for Opt where Opt is the cost of the optimal solution. But this gives $\$1614 - Opt = 0.076 \cdot Opt$ or $\$1614 = 0.076 \cdot Opt + Opt$. So,

$$Opt = \frac{\$1614}{1.076} = \$1500.$$

53. (a)

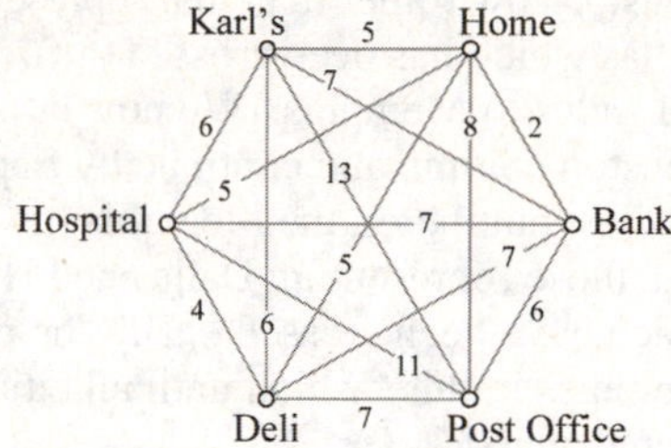

(b) Just "eyeballing it" will give the optimal tour in this case. It is: Home, Bank, Post Office, Deli, Hospital, Karl's , Home. The total length of the tour is 2 + 6 + 7 + 4 + 6 + 5 = 30 miles.

55. (a) Treat the guests as the vertices and let the friendships be represented by edges.

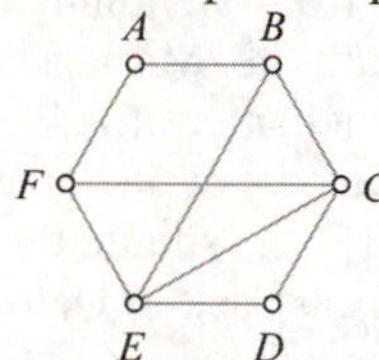

(b) Any Hamilton circuit gives a possible seating arrangement. One possibility is *A*, *B*, *C*, *D*, *E*, *F*.

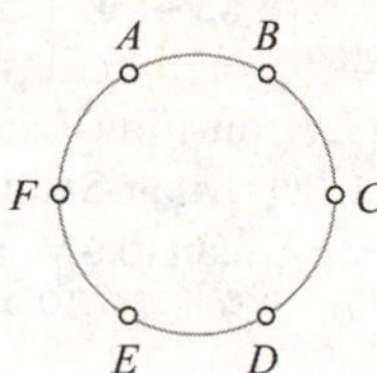

(c) Yes, there is. It corresponds to the Hamilton circuit *A*, *B*, *E*, *D*, *C*, *F*, *A*.

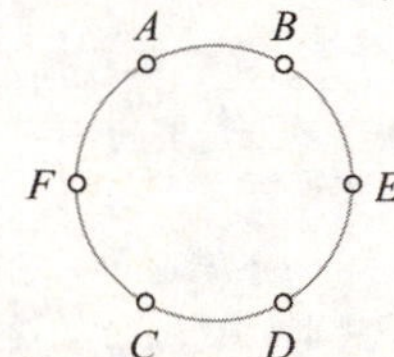

57. If we draw the graph describing the friendships among the guests (see figure) we can see that the graph does not have a Hamilton circuit, which means it is impossible to seat everyone around the table with friends on both sides.

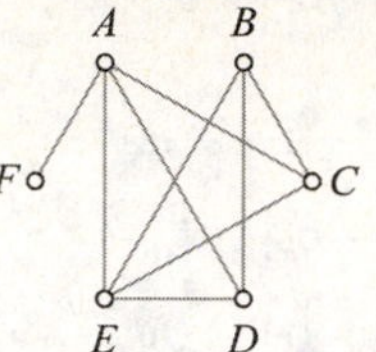

JOGGING

59.

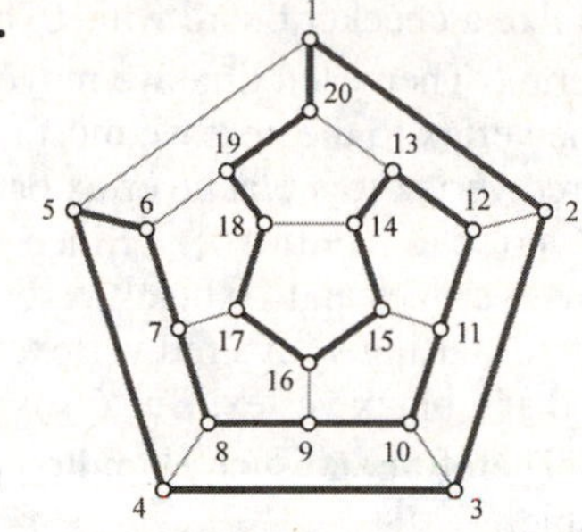

61. (a) 1001; K_{501} has 500 more edges than K_{500} (the difference between these two graphs is that in K_{501} the 501st vertex has an edge extending to each of the other 500 vertices). Similarly K_{502} has 501 more edges than K_{501}. So, K_{502} has 500 + 501 = 1001 more edges than K_{500}.

(b) 500 + 501 + 502 = 1503; see part (a)

63. The 2 by 2 grid graph cannot have a Hamilton circuit because each of the 4 corner vertices as well as the interior vertex *I* must be preceded and followed by a boundary vertex. But there are only 4 boundary vertices–not enough to go around.

65. (a)

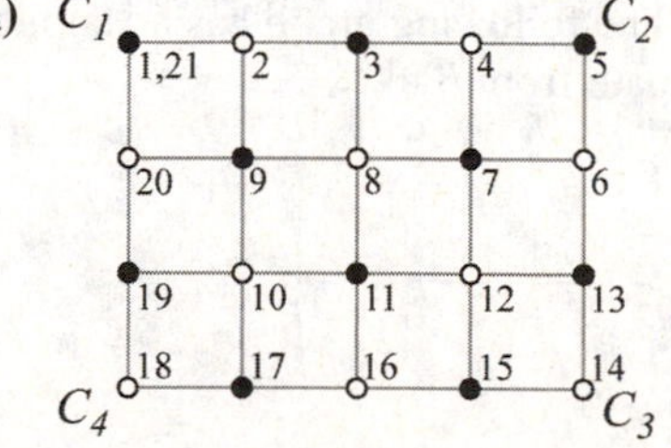

(b)

(c) Think of the vertices of the graph as being colored like a checker board with C_1 being a red vertex. Then each time we move from one vertex to the next we must move from a red vertex to a black vertex or from a black vertex to a red vertex. Since there are 10 red vertices and 10 black vertices and we are starting with a red vertex, we must end at a black vertex. But C_2 is a red vertex. Therefore, no such Hamilton path is possible.

67.

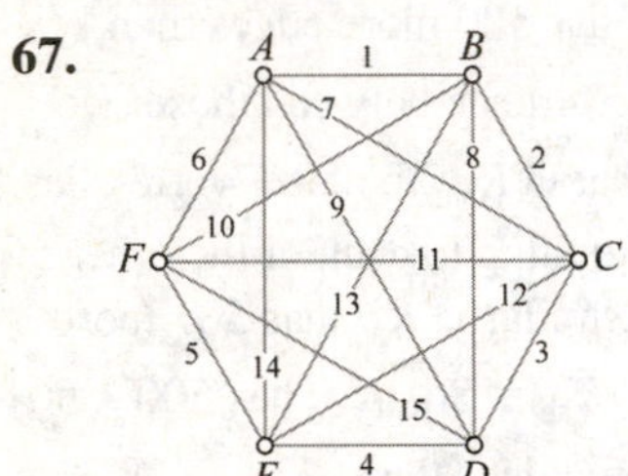

The optimal tour is *A*, *B*, *C*, *D*, *E*, *F*, *A*. The edges in this tour are those with the six lowest weights.

69. (a) Any circuit would need to cross the bridge twice (in order to return to where it started). But then each vertex at the end of the bridge would be included twice.

(b) The following graph has a Hamilton path from *B* to *C*.

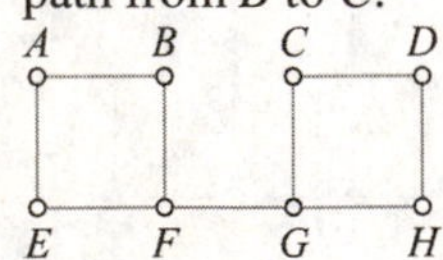

71. Dallas, Houston, Memphis, Louisville, Columbus, Chicago, Kansas City, Denver, Atlanta, Buffalo, Boston, Dallas.

The process starts by finding the smallest number in the Dallas row. This is 243 miles to Houston. We cross out the Dallas column and proceed by finding the smallest number in the Houston row (other than that representing Dallas which has been crossed out). This is 561 miles to Memphis. We now cross out the Houston column and continue by finding the smallest number in the Memphis row (other than those representing Dallas and Houston which have been crossed out). The process continues in this fashion until all cities have been reached.

73. (a) Julie should fly to Detroit. The optimal route will then be to drive 68 + 56 + 68 + 233 + 78 + 164 + 67 + 55 = 789 miles via Detroit-Flint-Lansing-Grand Rapids-Cheboygan-Sault Ste. Marie-Marquette-Escanaba-Menominee for a total cost of 789 × \$0.39 + \$3.00 = \$310.71. Since Julie can drive from Menominee back to Detroit (via Sault Ste. Marie and Cheboygan) in a matter of 227 + 78 + 280 = 585 miles at a cost of 585 × (\$0.39) + \$3.00 = \$231.15, she should do so and drop the rental car back in Detroit (assuming she need not pay extra for gas!). The total cost of her trip would then be \$310.71 + \$231.15 = \$541.86.

(b) The optimal route would be for Julie to fly to Detroit and drive 68 + 56 + 68 + 233 + 78 + 164 + 67 + 55 = 789 miles along the Hamilton path Detroit-Flint-Lansing-Grand Rapids-Cheboygan-Sault Ste. Marie-Marquette-Escanaba-Menominee for a total cost of 789 × \$0.49 + \$3.00 = \$389.61.

Chapter 7

WALKING

A. Trees

1. (a) Tree (this graph is a network, i.e. a connected graph, with no circuits)

(b) Not a tree (this graph is not connected - it has two components)

(c) Not a tree (this graph has a circuit *B,C,D,E,B*)

(d) Tree

3. (a) 7; see tree property 3

(b) 7; use your answer to (a) and tree property 2

5. (a) (II) Tree property 3 tells us that a tree with 8 vertices must have exactly 7 edges.

(b) (III)

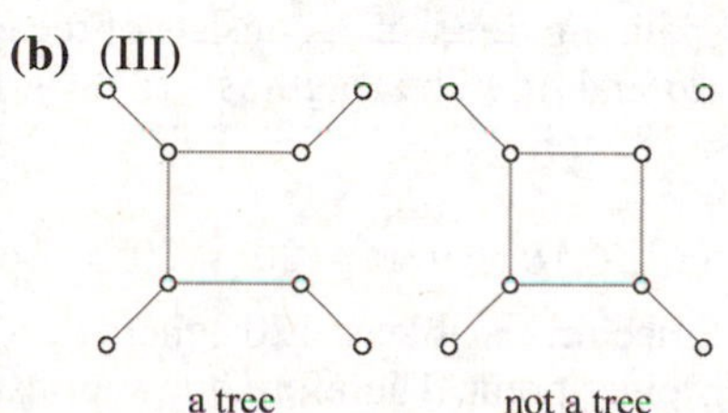

a tree　　not a tree

(c) (I) Tree property 2 says that a graph with 8 vertices in which every edge is a bridge must be a tree.

(d) (I) Tree property 1 says that if there is exactly one path joining any two vertices of a graph, the graph must be a tree.

(e) (I) Tree property 2 says that if every edge of a graph is a bridge, then the graph must be a tree.

7. (a) (III)

a tree　　not a tree

(b) (II) A tree has no circuits.

(c) (I) A graph with 8 vertices, 7 edges, and no circuits, must also be connected and hence must be a tree.

9. Since the degree of every vertex is even, the graph would have an Euler circuit (see chapter 5). Any graph with a circuit is not a tree.

B. Spanning Trees

11. (a) There are many spanning trees for this network. Below is one.

(b) Since this network has $N = 5$ vertices and $M = 10$ edges, the redundancy is $R = M - (N - 1) = 10 - (5 - 1) = 6$.

13. (a) There is only one spanning tree for this network since it is a tree. The spanning tree is the tree itself.

(b) Since this network has $N = 7$ vertices and $M = 6$ edges, the redundancy is $R = M - (N - 1) = 6 - (7 - 1) = 0$.

15. (a) Any one of the three edges *EI*, *DI*, or *DE* could be deleted to form a spanning tree.

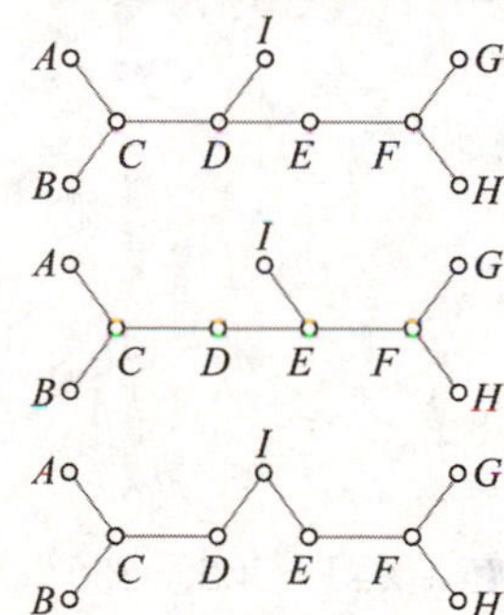

(b) Any one of the four edges *DJ*, *EJ*, *EI*, or *DI* could be deleted to form a spanning tree.

(c) 6; Any one of the edges *DI*, *IJ*, *JE*, *EK*, *KL*, or *LD* could be deleted to form a spanning tree.

17. (a) Each spanning tree excludes one of the edges *AB*, *BC*, *CA* and one of the edges *DI*, *IE*, *EJ*, *JD* so there are $3 \times 4 = 12$ different spanning trees.

(b) Each spanning tree excludes one of the edges *AB*, *BC*, *CA*, one of the edges *DI*, *IJ*, *JE*, *EK*, *KL*, *LD*, and one of the edges *FG*, *GH*, *HF*, so there are $3 \times 6 \times 3 = 54$ different spanning trees.

C. Minimum Spanning Trees and Kruskal's Algorithm

19. (a) Add edges to the tree in the following order: *EC*, *AD*, *AC*, *BC*.

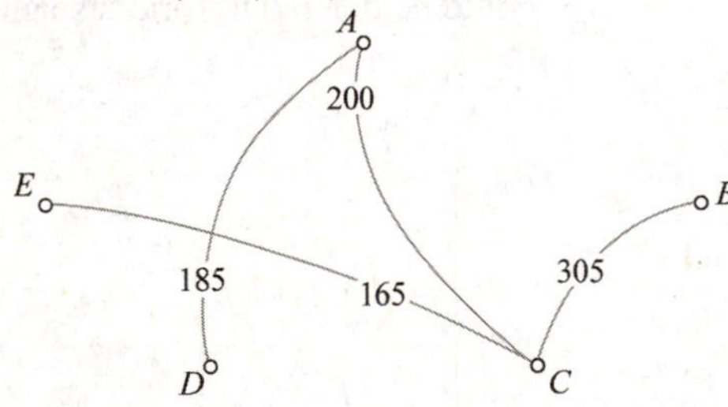

(b) 165 + 185 + 200 + 305 = 855

21. (a) Add edges to the tree in the following order: *DC*, *EF*, *EC*, *AB*, *AC*.

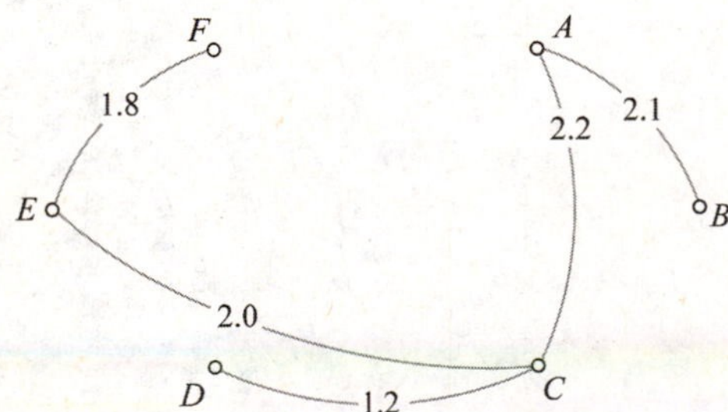

(b) $1.2 + 1.8 + 2.0 + 2.1 + 2.2 = 9.3$

23. Add edges to the tree in the following order: Kansas City – Tulsa (248), Pierre – Minneapolis (394), Minneapolis – Kansas City (447), Atlanta – Columbus (533), Columbus – Kansas City (656).

25. A spanning tree for the 20 vertices will have exactly 19 edges and so the cost will be

$$19 \text{ edges} \times \frac{1 \text{ mile}}{2 \text{ edges}} \times \frac{\$40,000}{1 \text{ mile}} = \$380,000.$$

D. Steiner Points and Shortest Networks

27. (a) $m(\angle CAB) = 180° - 24° - 27° = 129°$.

Since this is at least 120°, there is no Steiner point. Therefore, the shortest network is the minimum spanning tree (MST) which consists of the edges *AB* and *AC*. The length is 270 mi + 310 mi = 580 mi.

(b) $m(\angle CAB) = 180° - 33° - 27° = 120°$.

Since this is at least 120° (in fact, exactly that), there is no Steiner point. Therefore, the shortest network is the minimum spanning tree which consists of the edges *AB* and *AC*. The length is 212 mi + 173 mi = 385 mi.

29. (a) $m(\angle CAB) = 180° - 20° - 20° = 140°$.

Since this is at least 120°, there is no Steiner point. Therefore, the shortest network is the MST. We can deduce that the length of side *AC* is 183 km because $m(\angle ABC) = m(\angle ACB)$ causing the lengths of the sides opposite those angles must be equal (the triangle is isosceles). The MST consists of the edges *AB* and *AC*, and the length is 183 km + 183 km = 366 km.

(b) Since side *AC* is shorter than side *AB*, we deduce that $m(ABC) < m(ACB) = 28°$.

So, $m(CAB) > 180° - 28° - 28° = 124°$.

Since this is at least 120°, there is no Steiner point. Therefore, the shortest network is the MST which consists of the edges *AB* and *AC*. (Edge *BC* must be longer than either *AB* or *AC* since it is opposite the largest angle.) Its length is 181 km + 153 km = 334 km.

31. Since $m(CAB) = 128°$, the MST consists of the edges *AB* and *AC*. Using the Pythagorean Theorem, we compute the length of *AB* to be

$\sqrt{50^2+100^2}=\sqrt{12{,}500}\approx 111.8$ miles. Since the length of *AC* is the same as *AB*, the MST is approximately twice the length of *AB*. That is, approximately 224 miles.

33. The sum of the distances from *Z* to *A*, *B*, and *C* is 232 miles, the sum of the distances from *X* to *A*, *B*, and *C* is 240 miles, and the sum of the distances from *Y* to *A*, *B*, and *C* is 243 miles. Since one of the points is the Steiner point and the Steiner point is the point that makes the shortest network, *Z* is the Steiner point.

35. **(a)** *CE* + *ED* + *EB* is larger since *CD* + *DB* is the shortest network connecting the cities *C*, *D*, and *B*.

(b) *CD* + *DB* is the shortest network connecting the cities *C*, *D*, and *B* since $m(\angle CDB)=120°$ and so the shortest network is the same as the MST.

(c) *CE* + *EB* is the shortest network connecting the cities *C*, *E*, and *B* since $m(\angle CEB)>120°$ and so the shortest network is the same as the MST.

E. Miscellaneous

37. **(a)** k = 5, 2, 1, 0 are all possible. The network could have no circuits (k=5),

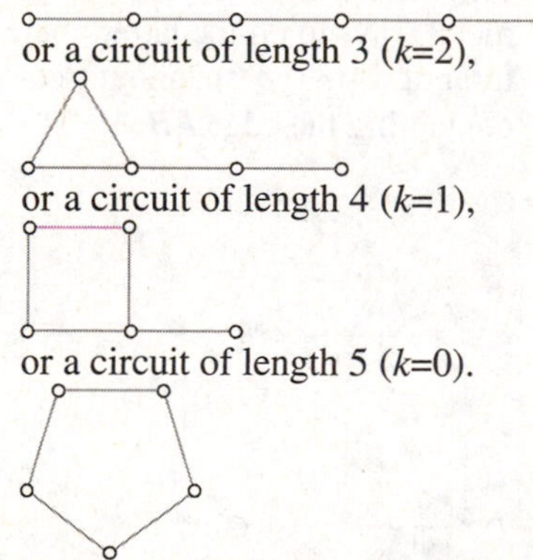

or a circuit of length 3 (k=2),

or a circuit of length 4 (k=1),

or a circuit of length 5 (k=0).

(b) Using the same pattern as in (a), values of $k=123$ and $0\le k\le 120$ are all possible.

39. **(a)** Since $M=N$ -1, the network is a tree by property 4. So, there are no circuits in the network.

(b) The redundancy of this network is R = 1 so one edge will need to be "discarded" in order to form a spanning tree. This means that there is one circuit in the network.

(c) 118. Each edge other than those in the circuit of length 5 is a bridge.

41. **(a)** Note that the graph is not connected.

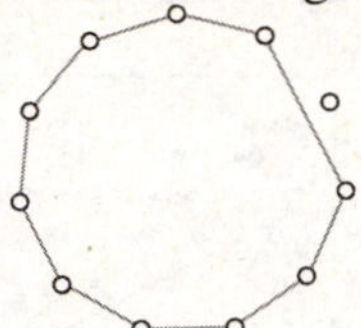

(b) The graph is, again, not connected.

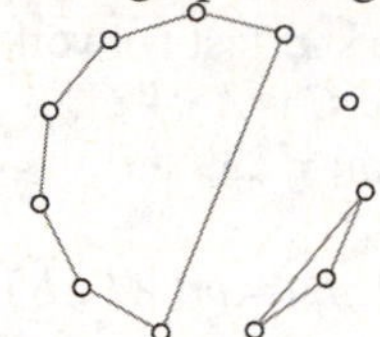

(c) Other examples are possible.

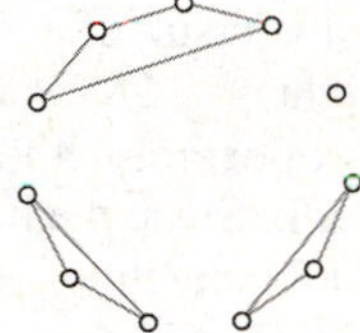

43. **(a)** Imagine that the figure is the top half of an equilateral triangle. Then, it becomes clear that h is twice the length of s. That is, $h=2s$.

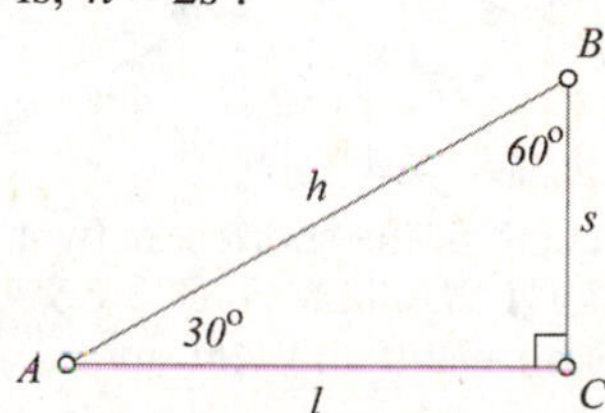

(b) By the Pythagorean theorem, $h^2=s^2+l^2$. Then, using the result of (a), we have $(2s)^2=s^2+l^2$ so that $4s^2=s^2+l^2$ and $3s^2=l^2$. Since $l>0$, it follows that $l=\sqrt{3}s$.

(c) $h=2s=2(21)=42$

$l=\sqrt{3}s=\sqrt{3}(21)\approx 36.37$

(d) $s=\frac{1}{2}h=\frac{1}{2}(30.4)=15.2$

$l=\sqrt{3}s=\sqrt{3}(15.2)\approx 26.30$

45. $m(\angle ASC)=m(\angle ASB)=m(\angle CSB)=120°$ since *S* is a Steiner point. Since *AB* = *AC*, $\triangle CSB$ is made up of two side by side 30°-60°-

90° triangles. Using the result of exercise 43(a), we have $CS = BS = 104$ mi. Since the height of $\triangle ABC$ is 83 mi, and the height of $\triangle CSB$ is 52 mi, it follows that $AS = 31$ mi.

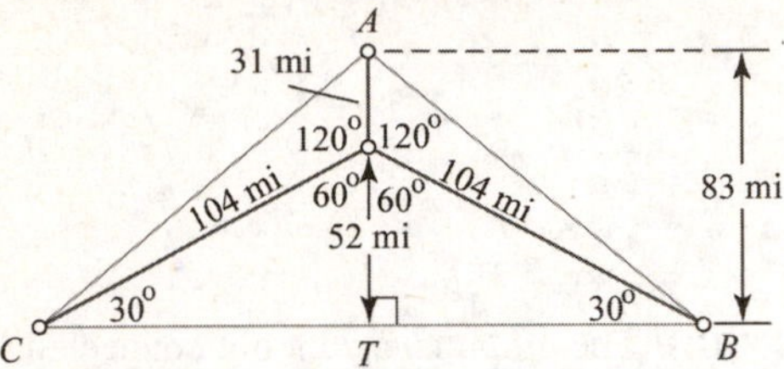

So, the length of the shortest network connecting A, B, and C is $AS + CS + BS = 31$ mi + 104 mi + 104 mi = 239 mi.

47. $m(\angle ASC) = m(\angle ASB) = m(\angle CSB) = 120°$ since S is a Steiner point. Since $AB = AC$, $\triangle CSB$ is made up of two side by side 30°-60°-90° triangles. Using the result of exercise 43(b), we see the height of $\triangle CSB$ is 100 mi. Then using the result of exercise 43(a), we have $CS = BS = 200$ mi. Since the height of $\triangle ABC$ is 150 mi, and the height of $\triangle CSB$ is 100 mi, it follows that $AS = 50$ mi.

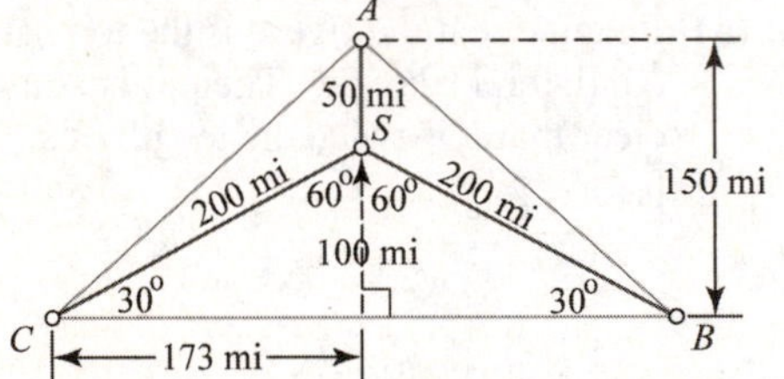

So, the length of the shortest network connecting A, B, and C is $AS + CS + BS = 50$ mi + 200 mi + 200 mi = 450 mi.

49. In a 30°-60°-90° triangle, the side opposite the 30° angle is one half the length of the hypotenuse and the side opposite the 60° angle is $\sqrt{3}/2$ times the length of the hypotenuse (see exercise 43). Therefore, the distance from A to J is $\sqrt{3}/2 \times 500$ miles $\approx$ 433 miles and so the total length of the T-network (rounded to the nearest mile) is 433 miles + 500 miles = 933 miles.

JOGGING

51. **(a)** 11; The key is to consider two cases: either EB is part of a spanning tree or it isn't. There are 5 spanning trees that do not contain edge EB. They can be found by removing EB along with one of the 5 edges of the circuit A,B,C,D,E,A.

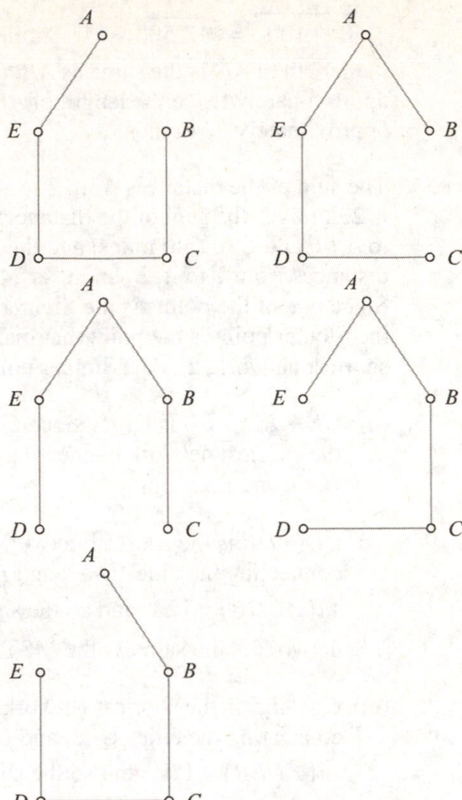

There are 6 spanning trees that do contain edge EB. They can be found by removing one of the 2 edges (not EB) of the circuit A,B,E,A along with one of the 3 edges (again, not EB) of the circuit B,C,D,E,B. Since there are 2 ways to do the former and 3 ways to do the latter, there are a total of $2 \times 3 = 6$ spanning trees containing the edge EB.

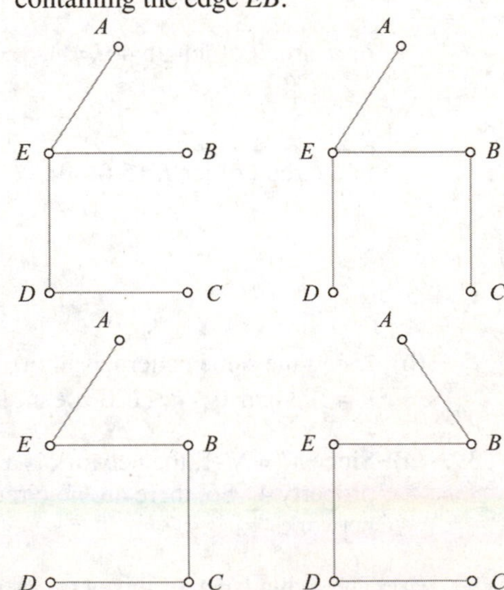

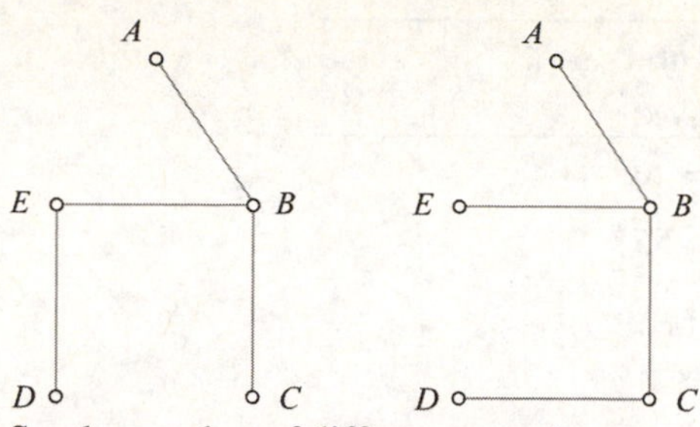

So, the number of different spanning trees is 5 + 6 = 11.

(b) 29; There are 9 spanning trees that do not contain edge *EF*. They can be found by removing *EF* along with one of the 9 edges of the circuit *A,B,C,D,E,I,H,G,F,A*. There are 20 spanning trees that do contain edge *EF*. They can be found by removing one of the 5 edges (not *EF*) of the circuit *A,B,C,D,E,F,A* along with one of the 4 edges (again, not *EF*) of the circuit *E,I,H,G,F,E*. There are 5 ways to do the former and 4 ways to do the latter making a total of $4\times5=20$ spanning trees containing the edge *EF*. Consequently there are 9 + 20 = 29 different spanning trees.

(c) 56; There are 8 spanning trees that do not contain edges *BG* and *CF*. They can be found by removing *BG* and *CF* along with one of the 8 edges of the circuit *A,B,C,D,E,F,G,H,A*. There are 15 spanning trees that contain edge *BG* but not edge *CF*. They can be found by removing one of the 3 edges (not *BG* of course) of the circuit *A,B,G,H,A* along with one of the 5 edges (again, not *BG*) of the circuit *B,C,D,E,F,G,B*. There are 3 ways to do the former and 5 ways to do the latter making a total of $3\times5=15$ spanning trees containing the edge *BG* but not the edge *CF*. Similarly, there are 15 spanning trees containing edge *CF* but not edge *BG*. Finally, there are 18 spanning trees that contain both the edges *BG* and *CF*. They can be found by removing one of the 3 edges (not *BG*) of the circuit *A,B,G,H,A* along with one of the 2 edges *BC* or *GF*, along with one of the 3 edges (not *CF*) of the circuit *C,D,E,F,C*. There are 3 ways to do the first, 2 ways to do the second and 3 ways to do the third making a total of $3\times2\times3=18$ such spanning trees. Consequently there are 8 + 15 + 15 + 18 = 56 different spanning trees.

53. (a)

(b)

(c)

(d)

55. If $M \geq N$, then the redundancy of the network is at least 1 and the network has at least one circuit. The shortest length of that circuit is 3. None of the edges in the circuit are bridges, but the remaining $M-3$ edges could all be bridges.

57.

Edges in Spanning Tree	Junction Points	Total Cost
AB, AC, AD	*A*	85 + 50 + 45 + 25 = 205
AB, BC, BD	*B*	85 + 90 + 75 + 5 = 255
AC, BC, CD	*C*	50 + 90 + 70 + 15 = 225
AD, BD, CD	*D*	45 + 75 + 70 + 20 = 210
AD, AB, BC	*A, B*	45 + 85 + 90 + 25 + 5 = 250
AB, AC, BD	*A, B*	85 + 50 + 75 + 25 + 5 = 240
AB, BC, CD	*B, C*	85 + 90 + 70 + 5 + 15 = 265
BD, BC, AC	*B, C*	75 + 90 + 50 + 5 + 15 = 235
BC, CD, AD	*C, D*	90 + 70 + 45 + 15 + 20 = 240
AC, CD, BD	*C, D*	50 + 70 +75 + 15 + 20 = 230
AB, AD, CD	*A, D*	85 + 45 + 70 + 25 + 20 = 245
AC, AD, BD	*A, D*	50 + 45 + 75 + 25 + 20 = 215
AD, AC, BC	*A, C*	45 + 50 + 90 + 25 + 15 = 225
AB, BD, CD	*B, D*	85 + 75 + 70 + 5 + 20 = 255
AD, BD, BC	*B, D*	45 + 75 + 90 + 20 + 5 = 235
AB, AC, CD	*A, C*	85 + 50 + 70 + 25 + 15 = 245

The minimum cost network connecting the 4 cities has a 3-way junction point at *A* and has a total cost of 205 million dollars.

59. (a) The switching station should be located 1 mile north of the airport. [Most of the options can be eliminated by common sense. Since there are only a handful of viable options, trial and error will discover this solution.]

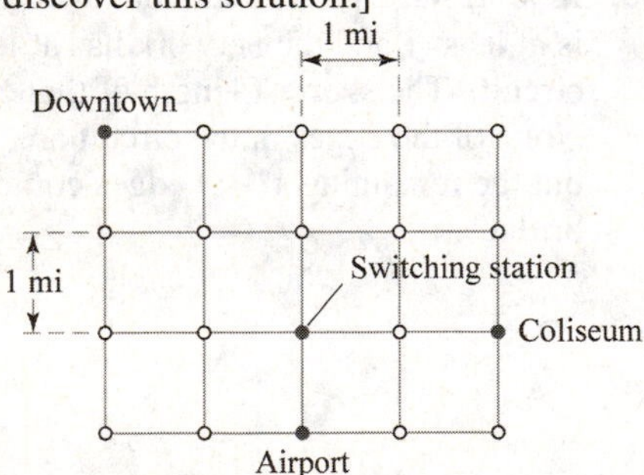

(b) Since the optimal network requires 7 miles of track and 1 switching station, the cost of the network is \$7.5 million.

61. (a) $m(\angle BFA) = 120^\circ$ since $m(\angle EFG) = 60^\circ$ (supplementary angles). Similarly, $m(\angle AEC) = 120^\circ$ and $m(\angle CGB) = 120^\circ$.

(b) In ΔABF, $m(\angle BFA) = 120^\circ$ [from part (a)] and so $m(\angle BAF) + m(\angle ABF) = 60^\circ$. It follows therefore that $m(\angle BAF) < 60^\circ$ and $m(\angle ABF) < 60^\circ$. Similarly, $m(\angle ACE) < 60^\circ$, $m(\angle CAE) < 60^\circ$, $m(\angle BCG) < 60^\circ$, and $m(\angle CBG) < 60^\circ$. Consequently, $m(\angle A) = m(BAF) + m(CAE) < 60^\circ + 60^\circ = 120^\circ$. Likewise, $m(\angle B) < 120^\circ$ and $m(\angle C) < 120^\circ$. So, triangle *ABC* has a Steiner point *S*.

(c) Any point X inside or on ΔABF (except vertex F) will have $m(\angle AXB) > 120^\circ$. Any point X inside or on ΔACE (except vertex E) will have $m(\angle AXC) > 120^\circ$. Any point X inside or on ΔBCG (except vertex G) will have $m(\angle BXC) > 120^\circ$. If S is the Steiner point, $m(\angle ASB) = m(\angle ASC) = m(BSC) = 120^\circ$, and so S cannot be inside or on ΔABF or ΔACE or ΔBCG. It follows that the Steiner point S must lie inside ΔEFG.

63. (a) The length of the network is $4x + (300 - x) = 3x + 300$ (see the figure which uses the results of exercise 43). The triangle formed by A, S_1, and the midpoint of AB is a 30°-60°-90° triangle. Therefore, $x = \frac{2}{\sqrt{3}} \times 200 = \frac{400\sqrt{3}}{3}$, and so the length of the network is $3\left(\frac{400\sqrt{3}}{3}\right) + 300 = 400\sqrt{3} + 300 \approx 993$ miles.)

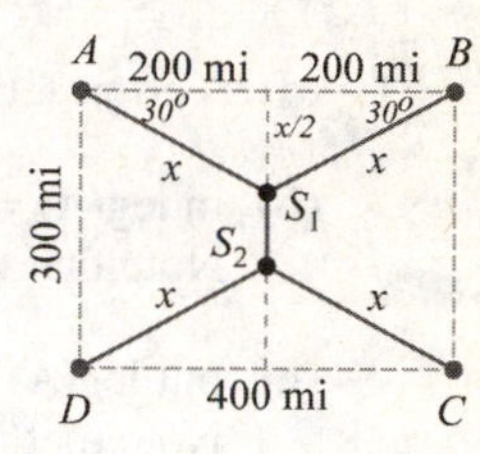

(b) The length of the network is $4x + (400 - x) = 3x + 400$, where $150^2 + \left(\frac{x}{2}\right)^2 = x^2$ (see figure). Solving gives $x = \frac{300\sqrt{3}}{3}$ and so the length of the network is $3\left(\frac{300\sqrt{3}}{3}\right) + 400 = 300\sqrt{3} + 400 \approx 919.6$ miles.

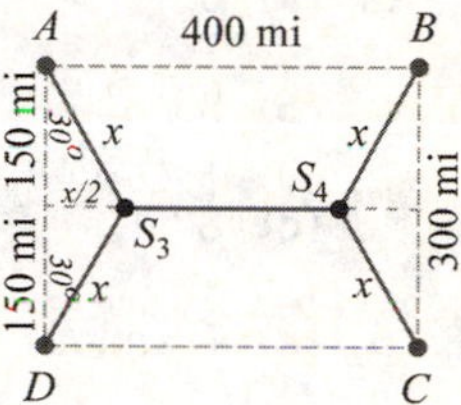

65. (a) There will be two subtrees formed as shown below.

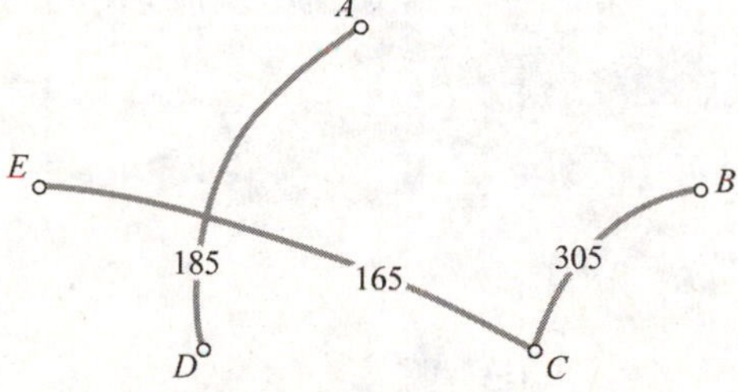

(b) The tree formed by Boruvka's algorithm is shown below.

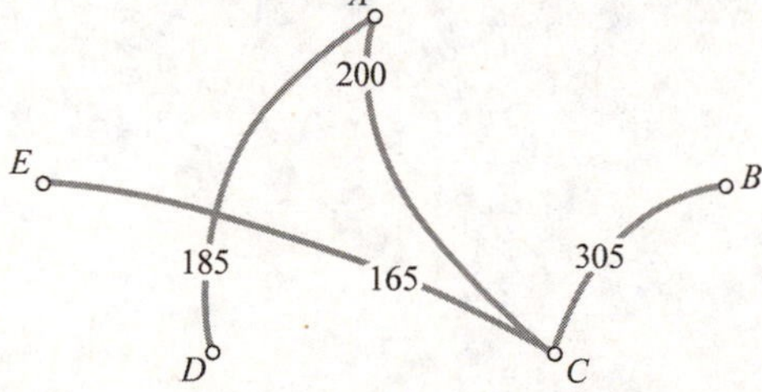

Chapter 8

WALKING

A. Directed Graphs

1. (a) Vertex set: V = {*A*, *B*, *C*, *D*, *E*}; Arc set: A = {*AB*, *AC*, *BD*, *CA*, *CD*, *CE*, *EA*, *ED*}

(b) indeg(*A*) = 2, indeg(*B*) = 1, indeg(*C*) = 1, indeg(*D*) = 3, indeg(*E*) = 1.
Note that the sum of the indegrees matches the number of arcs in A.

(c) outdeg(*A*) = 2, outdeg(*B*) = 1, outdeg(*C*) = 3, outdeg(*D*) = 0, outdeg(*E*) = 2.
Note that the sum of the outdegrees is the same as the sum of the indegrees and matches the number of arcs in A.

3. (a) 8

(b) 8

(c) 8

5. (a) *C* and *E*

(b) *B* and *C*

(c) *B*, *C*, and *E*

(d) No vertices are incident from *D*.

(e) *CD*, *CE* and *CA*

(f) No arcs are adjacent to *CD*.

7. (a)

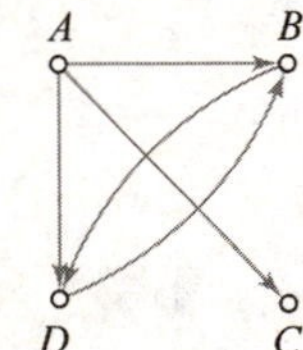

(b)

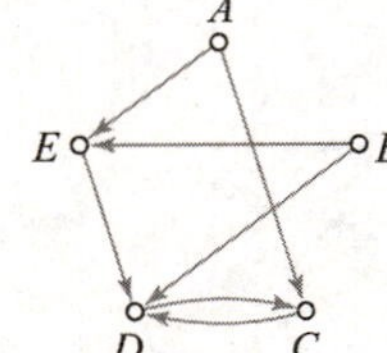

(c)

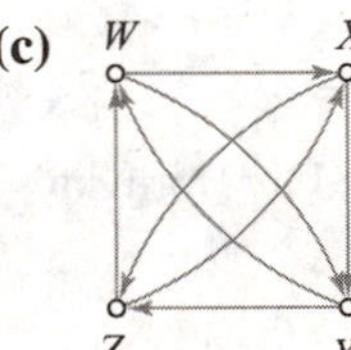

9. (a) 2 (these correspond to arcs *AB* and *AE*)

(b) 1 (this corresponds to arc *EA*)

(c) 1 (this corresponds to arc *DB*)

(d) 0 (there are no arcs incident to *D*)

11. (a) *A*, *B*, *D*, *E*, *F* is one possible path.

(b) *A*, *B*, *D*, *E*, *C*, *F*

(c) *B*, *D*, *E*, *B*

(d) The outdegree of vertex *F* is 0, so it cannot be part of a cycle.

(e) The indegree of vertex *A* is 0, so it cannot be part of a cycle.

(f) *B, D, E, B* is the only cycle

13.

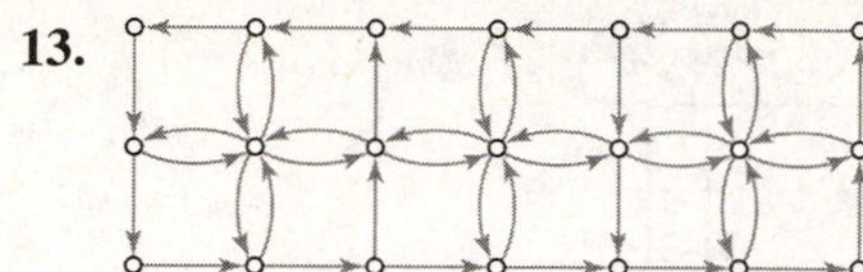

15. (a) *B*, since *B* is the only person that everyone respects.

(b) *A*, since *A* is the only person that no one respects.

(c) The individual corresponding to the vertex having the largest indegree (*B* in this example) would be the most reasonable choice for leader of the group since they would be the most respected person.

B. Project Digraphs

17. (a) V = {*START, A, B, C, D, E, F, G, H, END*};
A = {*START-A, START-H, AF, AC, B-END, CB, CE, D-END, E-END, FB, GB, GD, HG*}

(b)

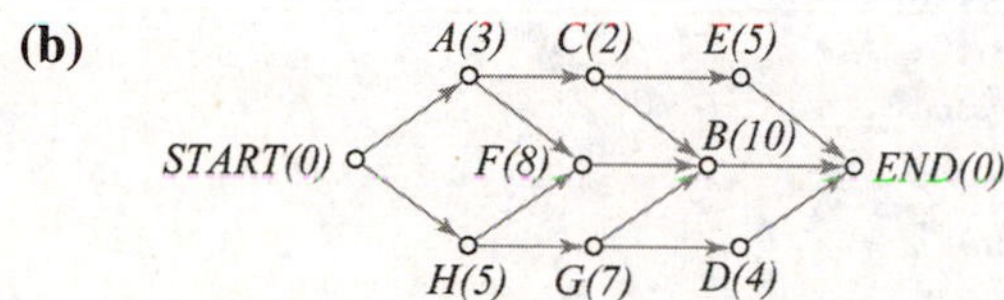

19. Vertex *A* represents the experiment taking 10 hours to complete. Vertices *B* and *C* represent the experiments requiring 7 hours each to complete. Vertices *D* and *E* represent the experiments requiring 12 hours each to complete. Vertices *F*, *G*, and *H* represent the experiments requiring 20 hours to complete.

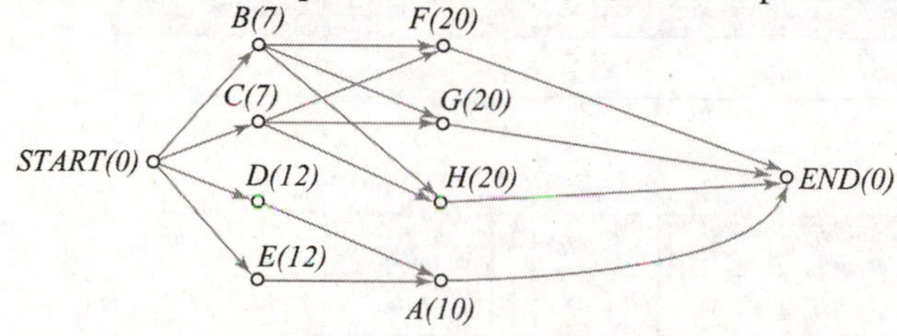

21.

L(1) P(32) K(12) G(8) W(4) C(4)

START(0) B(8) S(1) END(0)

F(1)

C. Schedules and Priority Lists

23. (a) There are 31 × 3 = 93 processor hours available. Of these, 10 + 7 + 11 + 8 + 9 + 5 + 3 + 6 + 4 + 7 + 5 = 75 hours are used. So, there must be 93 – 75 = 18 hours of idle time.

(b) There is a total of 75 hours of work to be done. Three processors working without any idle time would take $\frac{75}{3} = 25$ hours to complete the project.

25. There is a total of 75 hours of work to be done. Dividing the work equally between the six processors would require each processor to do $\frac{75}{6} = 12.5$ hours of work. Since there are no half hour jobs, the completion time could not be less than 13 hours.

27. 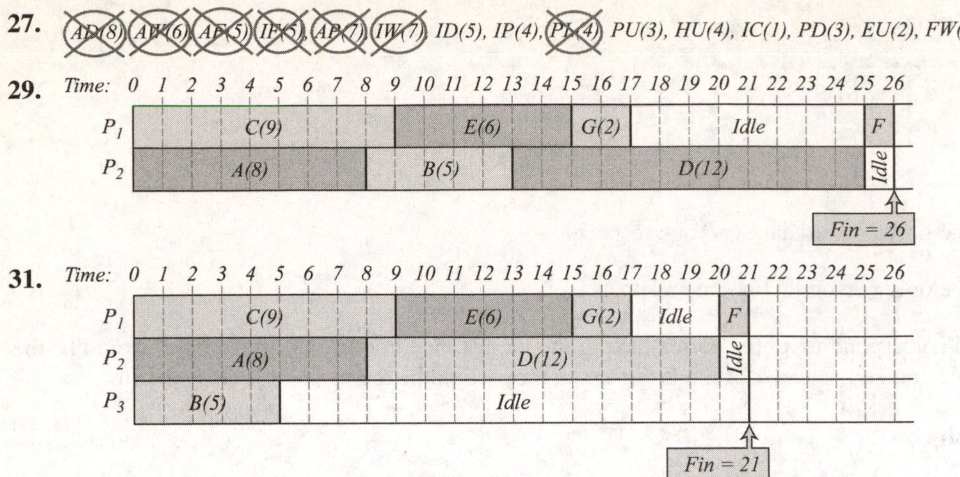

29.

31.

33. Both priority lists have all tasks in the top two paths of the digraph listed before any task in the bottom path.

35. The project digraph helps in the construction of the project timeline.

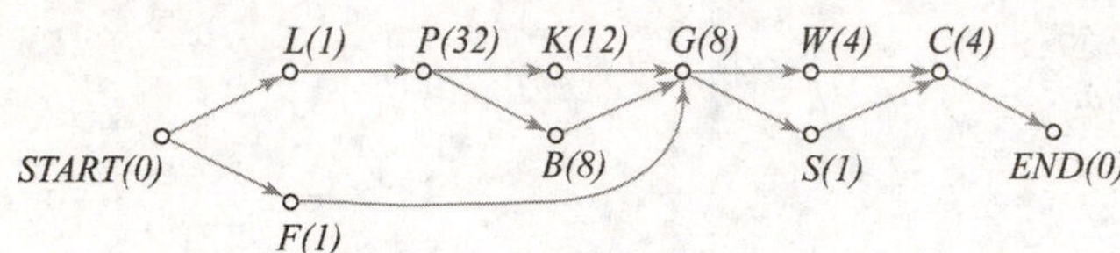

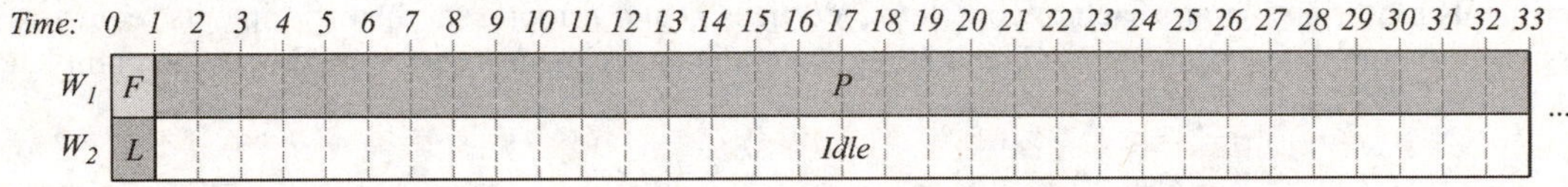

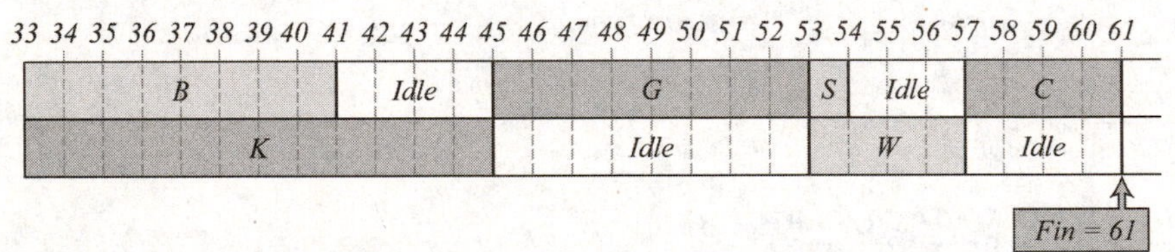

D. The Decreasing-Time Algorithm

37. Decreasing-Time List: *D(12), C(9), A(8), E(6), B(5), G(2), F(1)*

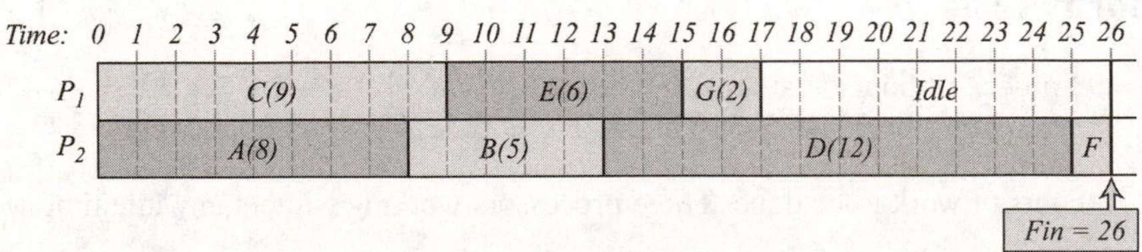

39. Decreasing-Time List: *K(20), D(15), I(15), C(10), E(7), H(5), J(5), B(4), G(4), F(3), A(2)*

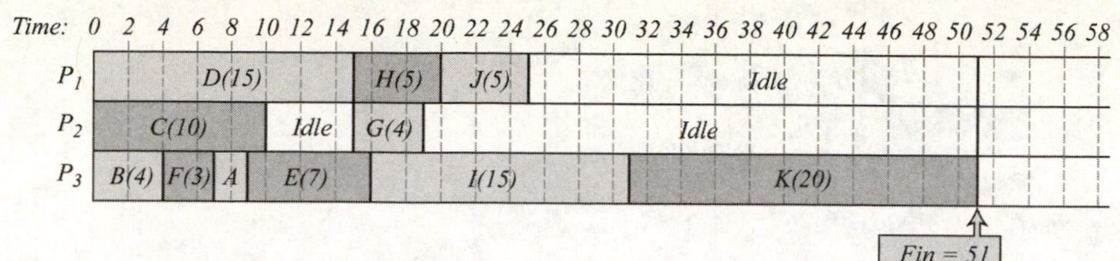

41. **(a)** Decreasing-Time List: *E(2), F(2), G(2), H(2), A(1), B(1), C(1), D(1)*

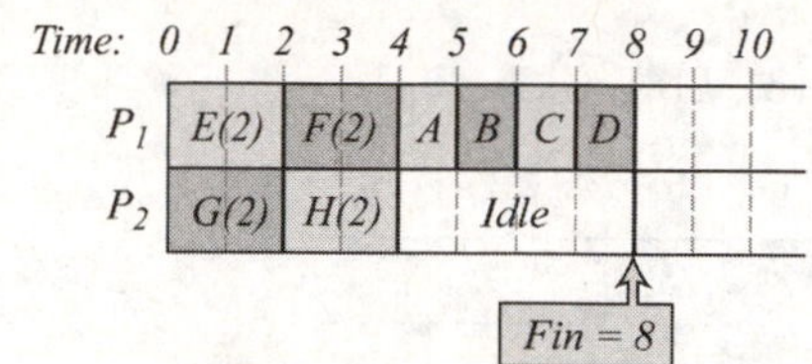

(b)

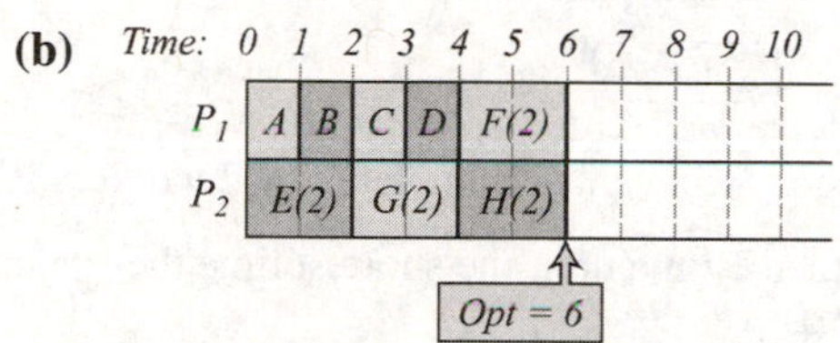

(c) The relative error expressed as a percent is $\varepsilon = \dfrac{Fin - Opt}{Opt} = \dfrac{8-6}{6} = 33\frac{1}{3}\%$.

43. **(a)** Decreasing-Time List: *E(5), F(5), G(5), H(5), I(5), J(5), A(4), B(4), C(4), D(4)*

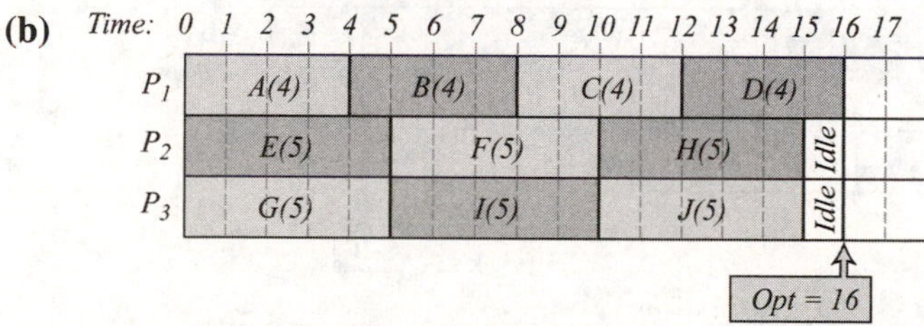

(b)

Time: 0 1 2 3 4 5 6 7 8 9 10 11 12 13 14 15 16 17

P1: A(4) | B(4) | C(4) | D(4)

P2: E(5) | F(5) | H(5) | Idle

P3: G(5) | I(5) | J(5) | Idle

Opt = 16

(c) The relative error expressed as a percent is $\varepsilon = \dfrac{Fin - Opt}{Opt} = \dfrac{26-16}{16} = 62.5\%$.

E. Critical Paths and the Critical-Path Algorithm

45. **(a)** As shown in the diagram, the critical time of each task is found by adding the times for each task in the shortest path formed between (and including) that task and *END*.

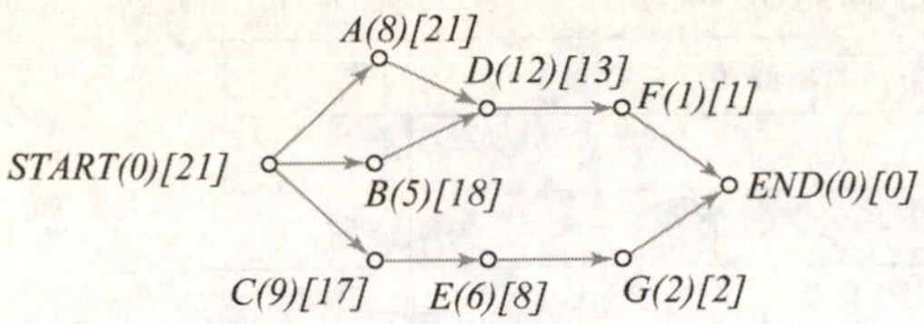

(b) The critical path is the longest time of a path between START and END. In this example, Start, *A*, *D*, *F*, END is the critical path, with a critical time of 21.

(c) From (a), the critical path list is: *A*[21], *B*[18], *C*[17], *D*[13], *E*[8], *G*[2], *F*[1].

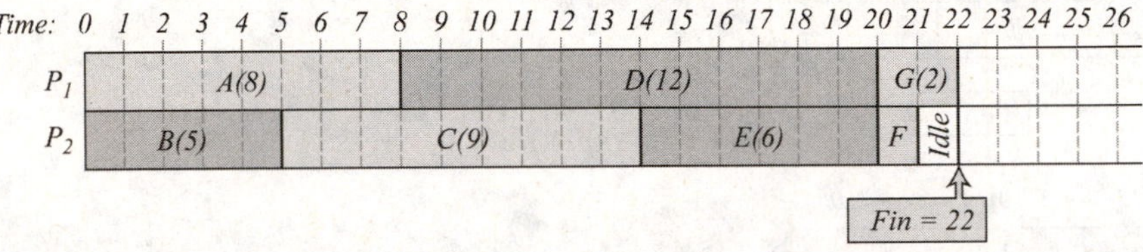

(d) There are a total of 43 work units, so the shortest time the project can be completed by 2 workers is $\frac{43}{2} = 21.5$ time units. Since there are no tasks with less than 1 time unit, the shortest time the project can actually be completed is 22 hours.

47. **(a)** The critical path list is: *B*[46], *A*[44], *E*[42], *D*[39], *F*[38], *I*[35], *C*[34], *G*[24], *K*[20], *H*[10], *J*[5]. So, the critical path is *START, B, E, I, K, END.*

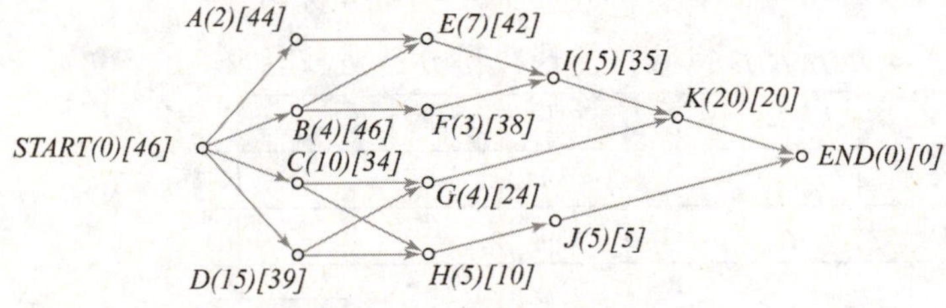

(b)

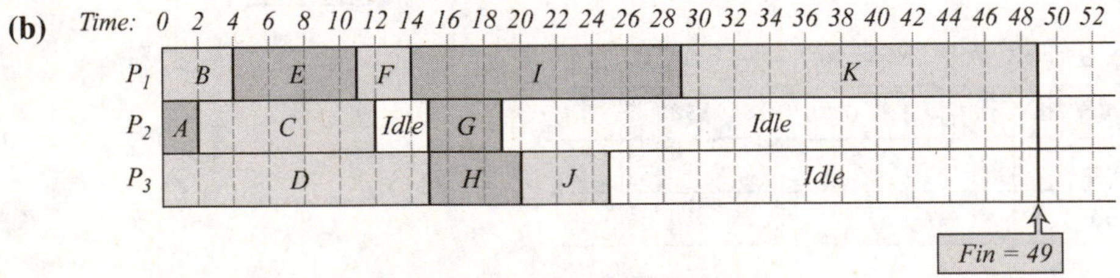

49. The project digraph, with critical times listed, is shown below.

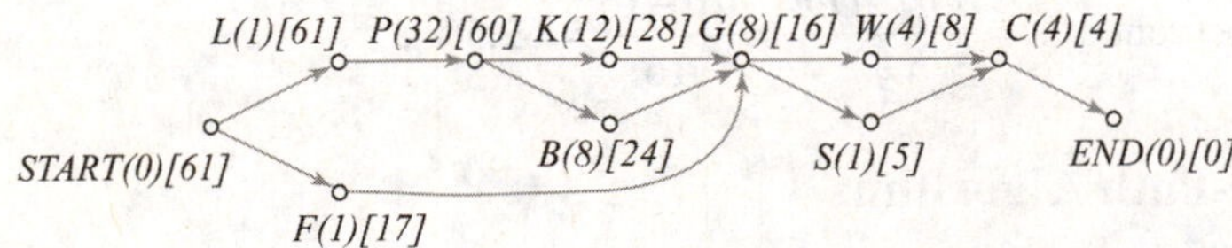

The critical-time priority list is *L[61], P[60], K[28], B[24], F[17], G[16], W[8], S[5], C[4]* and the critical path is *START[61], L[61], P[60], K[28], G[16], W[8], C[4], END[0]*.

Time: 0 3 6 9 12 15 18 21 24 27 30 33 36 39 42 45 48 51 54 57 60 63 66 69 72 75
P1
P2
P(32)
K(12)
G(8)
W
C
Idle
B(8)
Idle
Idle
Fin = 61

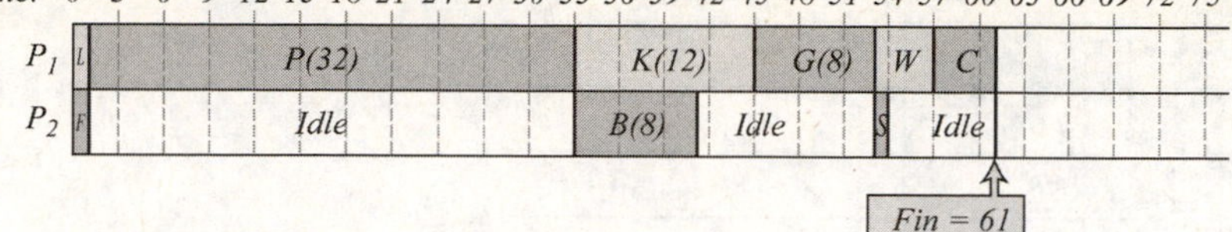

Since processing times are given in 15-minute units, the project finishing time is 15.25 hours.

F. Scheduling with Independent Tasks

51. (a)

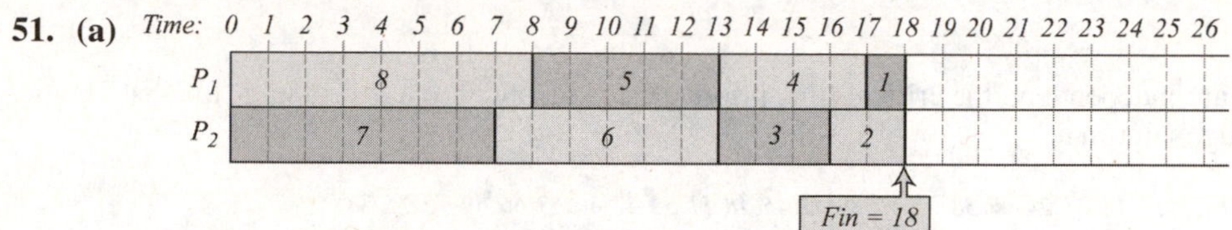

Time: 0 1 2 3 4 5 6 7 8 9 10 11 12 13 14 15 16 17 18 19 20 21 22 23 24 25 26
P1
P2
8
5
4
1
7
6
3
2
Fin = 18

(b) The optimal finishing time for $N = 2$ processors is $Opt = 18$ as shown in (a).

(c) The relative error expressed as a percent is $\varepsilon = \dfrac{Fin - Opt}{Opt} = \dfrac{18-18}{18} = 0\%$.

53. (a)

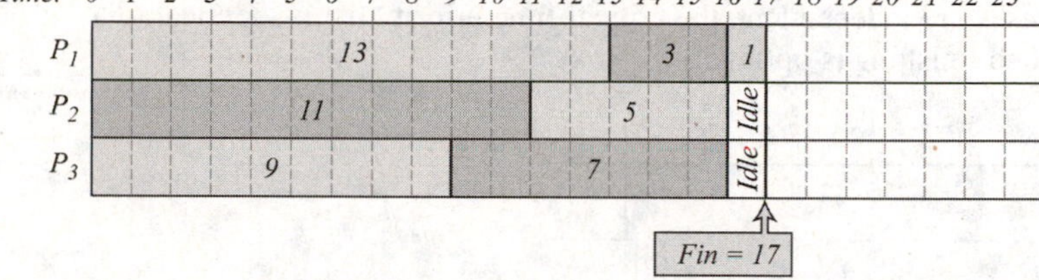

Time: 0 1 2 3 4 5 6 7 8 9 10 11 12 13 14 15 16 17 18 19 20 21 22 23
P1
P2
P3
13
3
1
11
5
Idle
9
7
Idle
Fin = 17

(b) The optimal finishing time for $N = 3$ processors is $Opt = 17$ as shown in (a). This is clear since the total length of all tasks is $1 + 3 + 5 + 7 + 9 + 11 + 13 = 49$ and $\dfrac{49}{3} > 16$.

(c) The relative error expressed as a percent is $\varepsilon = \dfrac{Fin - Opt}{Opt} = \dfrac{17-17}{17} = 0\%$.

55. (a) Since all tasks are independent, the critical-time priority list is identical to a decreasing-time list: *E[7], I[7], D[6], H[6], C[5], G[5], A[4], B[4], F[4]*.

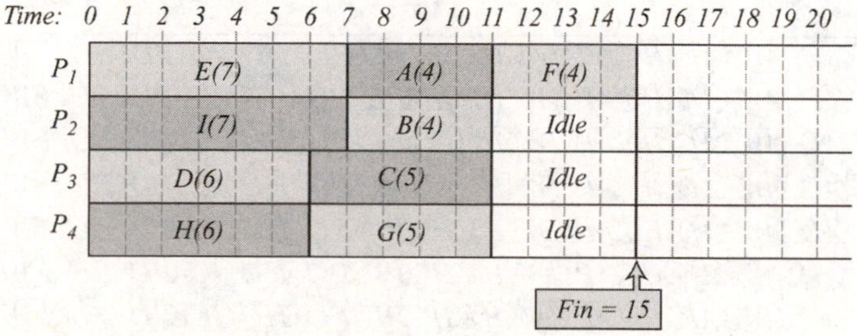

Time: 0 1 2 3 4 5 6 7 8 9 10 11 12 13 14 15 16 17 18 19 20
P1
P2
P3
P4
E(7)
A(4)
F(4)
I(7)
B(4)
Idle
D(6)
C(5)
Idle
H(6)
G(5)
Idle
Fin = 15

(b)

(c) The relative error expressed as a percent is $\varepsilon = \dfrac{Fin - Opt}{Opt} = \dfrac{15-12}{12} = 25\%$.

57. (a) Since all tasks are independent, the critical-time priority list is identical to a decreasing-time list. Notice that the finishing is optimal.

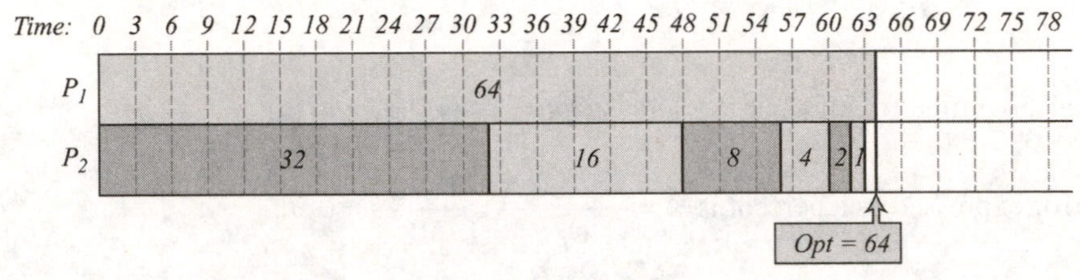

(b) Once again, since all tasks are independent, the critical-time priority list is identical to a decreasing-time list. The finishing is optimal.

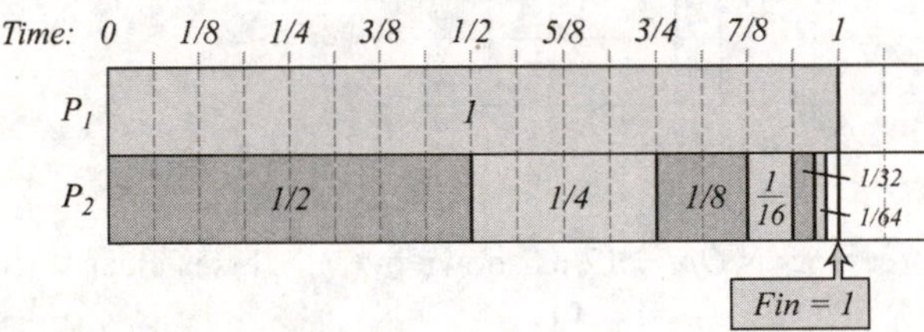

G. Miscellaneous

59. Every arc of the graph contributes 1 to the indegree sum and 1 to the outdegree sum.

61. Assuming that all of the tasks are independent, then *A, B, C, E, G, H, D, F, I* is one possible priority list.

63. Refer to the project digraph given in the text.

Time	Status of Processors		Priority List
	P_1	P_2	
20	Start: *IW*	Start: *PL*	~~AD~~ ~~AW~~ ~~AF~~ ~~IF~~ ~~AP~~ ~~IW~~ ID IP ~~PL~~ PU HU IC PD EU FW
24	Busy: *IW*	Idle	~~AD~~ ~~AW~~ ~~AF~~ ~~IF~~ ~~AP~~ ~~IW~~ ID IP ~~PL~~ PU HU IC PD EU FW
27	Start: *ID*	Start: *IP*	~~AD~~ ~~AW~~ ~~AF~~ ~~IF~~ ~~AP~~ ~~IW~~ ~~ID~~ ~~IP~~ ~~PL~~ PU HU IC PD EU FW
31	Busy: *ID*	Start: *HU*	~~AD~~ ~~AW~~ ~~AF~~ ~~IF~~ ~~AP~~ ~~IW~~ ~~ID~~ ~~IP~~ ~~PL~~ PU ~~HU~~ IC PD EU FW
32	Start: *PU*	Busy: *HU*	~~AD~~ ~~AW~~ ~~AF~~ ~~IF~~ ~~AP~~ ~~IW~~ ~~ID~~ ~~IP~~ ~~PL~~ ~~PU~~ ~~HU~~ IC PD EU FW
35	Start: *IC*	Start: *PD*	~~AD~~ ~~AW~~ ~~AF~~ ~~IF~~ ~~AP~~ ~~IW~~ ~~ID~~ ~~IP~~ ~~PL~~ ~~PU~~ ~~HU~~ ~~IC~~ ~~PD~~ ~~EU~~ ~~FW~~
36	Start: *EU*	Busy: *PD*	~~AD~~ ~~AW~~ ~~AF~~ ~~IF~~ ~~AP~~ ~~IW~~ ~~ID~~ ~~IP~~ ~~PL~~ ~~PU~~ ~~HU~~ ~~IC~~ ~~PD~~ ~~EU~~ ~~FW~~
38	Start: *FW*	Idle	~~AD~~ ~~AW~~ ~~AF~~ ~~IF~~ ~~AP~~ ~~IW~~ ~~ID~~ ~~IP~~ ~~PL~~ ~~PU~~ ~~HU~~ ~~IC~~ ~~PD~~ ~~EU~~ ~~FW~~
44	Idle	Idle	~~AD~~ ~~AW~~ ~~AF~~ ~~IF~~ ~~AP~~ ~~IW~~ ~~ID~~ ~~IP~~ ~~PL~~ ~~PU~~ ~~HU~~ ~~IC~~ ~~PD~~ ~~EU~~ ~~FW~~

Time: 0 2 4 6 8 10 12 14 16 18 20 22 24 26 28 30 32 34 36 38 40 42 44

P_1: AD | AP | IF | IW | ID | PU | IC | EU | FW

P_2: AW | AF | Idle | PL | Idle | IP | HU | PD | Idle

Fin = 44

65. Referring to the project digraph with critical times as given in the text, we see the critical path priority list is: *AP*[34], *AF*[32], *AW*[28], *IF*[27], *IW*[22], *AD*[18], *IP*[15], *PL*[11], *HU*[11], *ID*[10], *IC*[7], *FW*[6], *PU*[5], *PD*[3], *EU*[2].

Time	*Status of Processors*		*Priority List*
	P_1	P_2	
0	Start: AP	Start: AF	AP AF AW IF IW AD IP PL HU ID IC FW PU PD EU
5	Busy: AP	Start: AW	AP AF AW IF IW AD IP PL HU ID IC FW PU PD EU
7	Start: IF	Busy: AW	AP AF AW IF IW AD IP PL HU ID IC FW PU PD EU
11	Busy: IF	Start: AD	AP AF AW IF IW AD IP PL HU ID IC FW PU PD EU
12	Start: IW	Busy: AD	AP AF AW IF IW AD IP PL HU ID IC FW PU PD EU
19	Start: IP	Start: PL	AP AF AW IF IW AD IP PL HU ID IC FW PU PD EU
23	Start: HU	Start: ID	AP AF AW IF IW AD IP PL HU ID IC FW PU PD EU
27	Start: IC	Busy: ID	AP AF AW IF IW AD IP PL HU ID IC FW PU PD EU
28	Start: FW	Start: PU	AP AF AW IF IW AD IP PL HU ID IC FW PU PD EU
31	Busy: FW	Start: PD	AP AF AW IF IW AD IP PL HU ID IC FW PU PD EU
34	Start: EU	Idle	AP AF AW IF IW AD IP PL HU ID IC FW PU PD EU
36	Idle	Idle	AP AF AW IF IW AD IP PL HU ID IC FW PU PD EU

The schedule is:

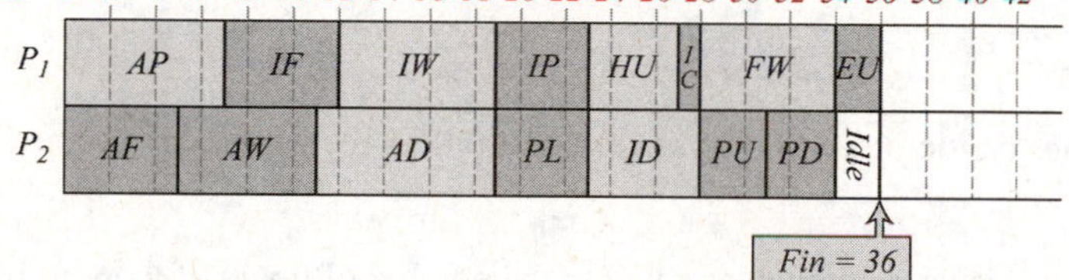

JOGGING

67. (a) When $N = 7$, $\varepsilon = \dfrac{7-1}{3\times 7} = \dfrac{6}{21} \approx 28.57\%$ so the range of errors is between 0 and 28.57%.

When $N = 8$, $\varepsilon = \dfrac{8-1}{3\times 8} = \dfrac{7}{24} \approx 29.17\%$ so the range of errors is between 0 and 29.17%.

When $N = 9$, $\varepsilon = \dfrac{9-1}{3\times 9} = \dfrac{8}{27} \approx 29.63\%$ so the range of errors is between 0 and 29.63%.

When $N = 10$, $\varepsilon = \dfrac{10-1}{3\times 10} = \dfrac{9}{30} = 30\%$ so the range of errors is between 0 and 30%.

(b) Since $N - 1 < N$, we have for every N that $\dfrac{N-1}{3N} \le \dfrac{N}{3N} = \dfrac{1}{3} = 33\dfrac{1}{3}\%$.

(c) As N gets larger, the difference between N and N-1 gets smaller (that is, (N-1)/N gets closer to 1).

Alternatively, since $\dfrac{N-1}{3N} = \dfrac{1-\frac{1}{N}}{3}$, the Graham bound gets closer to 1/3 as N gets larger.

69. **(a)**

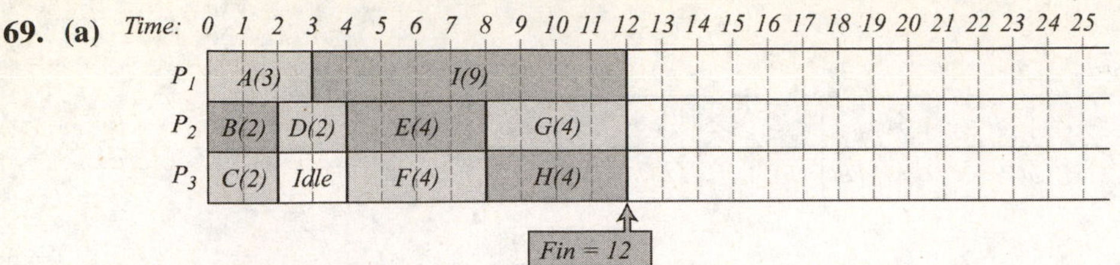

(b)

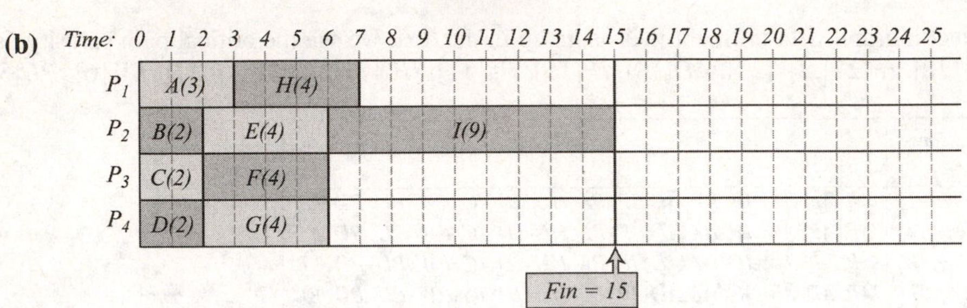

(c) An extra processor was used and yet the finishing time of the project increased.

71. **(a)** This is the same question and answer as 69(a).

(b)

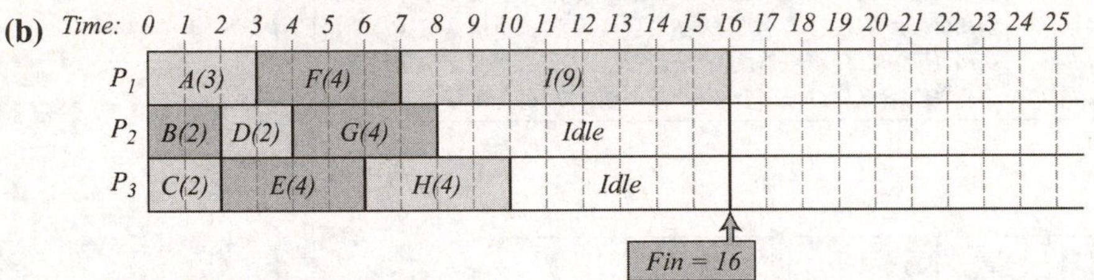

(c) The project had fewer restrictions on the order of the assignments to the processors and yet the finishing time of the project increased.

73. **(a)** The finishing time of a project is always more than or equal to the number of hours of work to be done divided by the number of processors doing the work.

(b) The schedule is optimal with no idle time.

(c) The total idle time in the schedule.

Mini-Excursion 2

A. Graph Colorings and Chromatic Numbers

1. (a) Many colorings are possible. One has vertices *A,G,C,J,E* colored blue, *B,H,K,F* colored red, and *D,I,L* colored green.

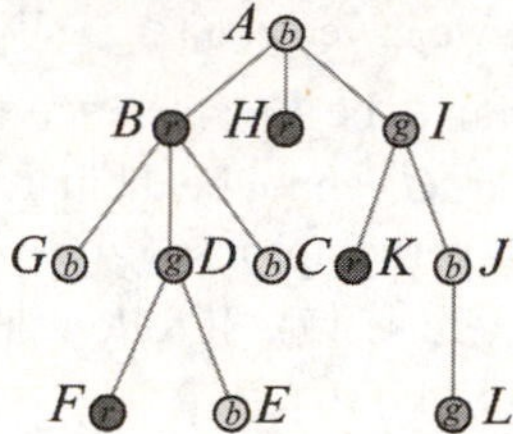

(b) One coloring has vertices *A,G,D,C,K,J* colored blue, and *B,H,I,F,E,L* colored red.

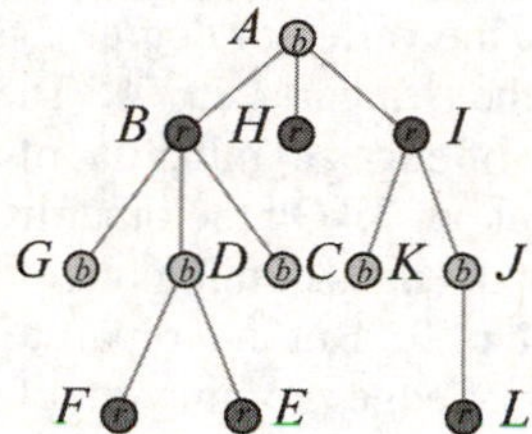

(c) $\chi(G) = 2$. At least 2 colors are needed and (b) shows that *G* can be colored with 2 colors.

3. (a) Many colorings are possible. One has vertices *A* and *C* colored blue, *B* and *E* colored red, *D* colored green, *G* colored yellow, and *F* colored purple.

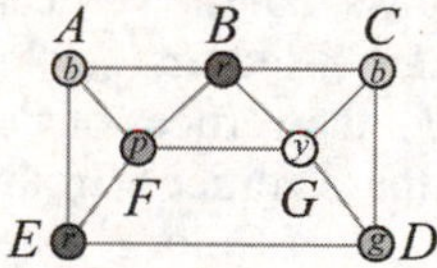

(b) One coloring has vertices *A* and *G* colored blue, *B* and *E* colored red, *C* and *F* colored green, and *D* colored yellow.

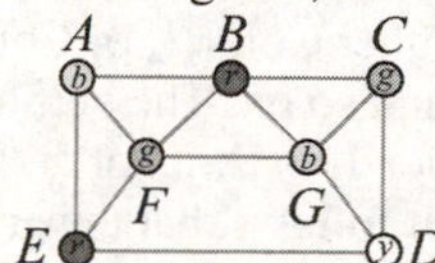

(c) $\chi(G) = 4$. The graph cannot be colored with 3 colors. If we try to color *G* with 3 colors and start with a triangle, say *AEF*, and color it blue, red, green, then *B* is forced to be red, *G* is forced to be blue, *C* is forced to be green, and then *D* will require a fourth color.

5. (a) One coloring has vertices *A,C,F* and *I* colored blue, *B,D* and *G* colored red, and *E,H* and *J* colored green.

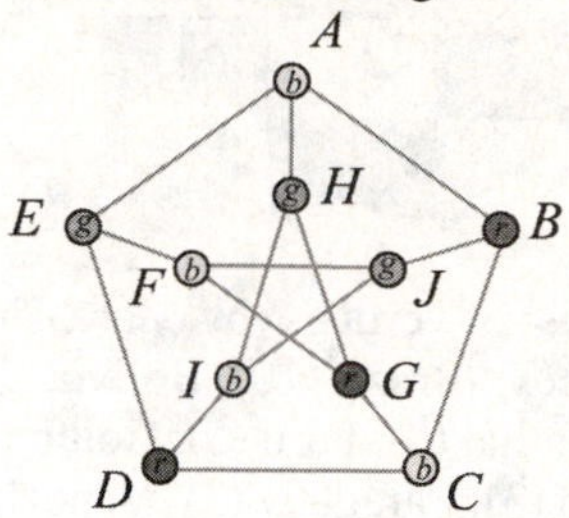

(b) $\chi(G) = 3$. Three colors are needed because *A*, *B*, *C*, *D*, and *E* form a circuit of length 5.

7. (a) Every vertex of K_n is adjacent to every other vertex, so every vertex has to be colored with a different color.

(b) $\chi(G) = n-1$. If we remove one edge from K_n, then there are two vertices that are not adjacent. They can be colored with the same color (say blue). The remaining vertices have to be colored with different colors other than blue.

9. (a) Adjacent vertices around the circuit can alternate colors (blue, red, blue, red, ...).

(b) To color an odd circuit we start by alternating two colors (blue, red, blue, red,...), but when we get to the last vertex, it is adjacent to both a blue and a red vertex, so a third color is needed.

11. $\chi(G) = 2$.

Since a tree has no circuits, we can start with any vertex *v*, color it blue and alternate blue, red, blue, red, blue, ... along any path of the tree. Every vertex of the tree is in a unique path joining it to *v* and can be colored either red or blue.

B. Map Coloring

13. Answers may vary. One possible coloring is shown below.

15. (a),(b) List of vertices (by decreasing order of degrees): Brazil (10), Bolivia (5), Argentina (5), Peru (5), Columbia (4), Chile (3), Paraguay (3), Venezuela (3), Guyana (3), Suriname (3), Ecuador (2), French Guiana (2), Uruguay (2).
Priority list of colors: Blue, Red, Green, Yellow.

Blue vertices: Chile, Ecuador, Brazil; Red vertices: Bolivia, Columbia, Guyana, French Guiana, Uruguay; Green vertices: Argentina, Peru, Venezuela, Suriname; Yellow vertex: Paraguay

(c) The chromatic number is 4 since Brazil, Bolivia, Argentina and Paraguay are all adjacent to each other.

C. Miscellaneous

17. (a) All the vertices in *A* can be colored with the same color, and all of the vertices in *B* can be colored with a second color.

(b) Suppose *G* has $\chi(G) = 2$. Say the vertices are colored blue and red. Then let *A* denote the set of blue vertices and *B* the set of red vertices.

(c) If *G* had a circuit with an odd number of vertices then that circuit alone would require 3 colors (see exercise 9). But from (a) we know that $\chi(G) = 2$.

19. (a) Since the sum of the degrees of all the vertices must be even (see Euler's Sum of Degrees theorem in chapter 5), it follows that the number of vertices of degree 3 must be even, and thus, *n* must be odd. If $n = 3$, there can be no vertices of degree 3 unless there are multiple edges.

(b) From the strong version of Brook's theorem we have $\chi(G) \leq 3$. We know $\chi(G)$ cannot be 1 (see exercise 8). Moreover, *G* cannot be a bipartite graph (see next paragraph), so $\chi(G)$ cannot be 2 (see exercise 17). It follows that $\chi(G) = 3$.

A graph with *n-1* vertices of degree 3 and one vertex of degree 2 cannot be bipartite because the vertex of degree 2 must be in one of the two parts, say *A*. Then the number of edges coming out of *A* is 2 plus a multiple of 3. On the other hand, the number of edges coming out of *B* is a multiple of 3. But in a bipartite graph the number of edges coming out of part *A* must equal the number of edges coming out of part *B* (they are both equal to the total number of edges in the graph).

21. Suppose that the graph *G* is colored with $\chi(G)$ colors: Color 1, Color 2, ..., Color *K* (for simplicity we will use *K* for $\chi(G)$). Now make a list $v_1, v_2, \ldots, v_n$ of the vertices of the graph as follows: All the vertices of Color 1 (in any order) are listed first (call these vertices Group 1), the vertices of Color 2 are listed next (call these vertices Group 2), and so on, with the vertices of Color *K* listed last (Group *K*). Now when we apply the greedy algorithm to this particular list, the vertices in Group 1 get Color 1, the vertices in Group 2 get either Color 1 or Color 2, the vertices in Group 3 get either Color 1, or Color 2, or at worst, Color 3, and so on. The vertices in Group *K* get Color 1, or 2, ..., or at worst, Color *K*. It follows that the greedy algorithm gives us an optimal coloring of the graph.

23. The solution appears in mini-excursion 2 after the References and Further Readings.

Chapter 9

WALKING

A. Fibonacci Numbers

1. (a) $F_{10} = 55$ (the 10th Fibonacci number)

(b) $F_{10} + 2 = 55 + 2 = 57$

(c) $F_{10+2} = F_{12} = 144$

(d) $F_{10} / 2 = 55/2 = 27.5$

(e) $F_{10/2} = F_5 = 5$

3. (a) $F_1 + F_2 + F_3 + F_4 + F_5 = 1+1+2+3+5$
$= 12$

(b) $F_{1+2+3+4+5} = F_{15} = 610$

(c) $F_3 \times F_4 = 2 \times 3 = 6$

(d) $F_{3\times4} = F_{12} = 144$

(e) $F_{F_4} = F_3 = 2$

5. (a) $3F_N + 1$ represents one more than three times the *N*th Fibonacci number.

(b) $3F_{N+1}$ represents three times the Fibonacci number in position $(N + 1)$.

(c) $F_{3N} + 1$ represents one more than the Fibonacci number in the 3*N*th position.

(d) F_{3N+1} represents the Fibonacci number in position $(3N + 1)$.

7. (a) $F_{38} = F_{37} + F_{36}$
$= 24{,}157{,}817 + 14{,}930{,}352$
$= 39{,}088{,}169$

(b) $F_{35} = F_{37} - F_{36}$
$= 24{,}157{,}817 - 14{,}930{,}352$
$= 9{,}227{,}465$

9. (I) $F_{N+2} = F_{N+1} + F_N$, $N > 0$, is an equivalent way to express the fact that each term of the Fibonacci sequence is equal to the sum of the two preceding terms.

11. (a) 47 = 34 + 13
Note that 34 is the largest Fibonacci number less than 47.

(b) 48 = 34 + 13 + 1
34 is the largest Fibonacci number less than 48. Then, note that 13 is the largest Fibonacci number less than what remains (48 – 34 = 14).

(c) 207 = 144 + 55 + 8
144 is the largest Fibonacci number less than 207. Next, 55 is the largest Fibonacci number less than what remains (207 – 144 = 63).

(d) 210 = 144 + 55 + 8 + 3
144 is the largest Fibonacci number less than 210. Also, 55 is the largest Fibonacci number less than the remaining 210 – 144 = 66. Finally, 8 is the largest Fibonacci number less than the remaining 210 – 144 – 55 = 11.

13. (a) Fifth equation in the sequence:
1 + 2 + 5 + 13 + 34 + 89 = 144
(In words, the left hand side is the sum of every other Fibonacci number. The right hand side is the next Fibonacci after the last one appearing on the left hand side.)

(b) 22. In each case the Fibonacci number that appears on the right of the equality is that Fibonacci number immediately following the largest Fibonacci number appearing on the left.

(c) *N*+1. *N*+1 is one bigger than *N*.

15. (a) Choosing, for example, the Fibonacci numbers $F_4 = 3$, $F_5 = 5$, $F_6 = 8$, and $F_7 = 13$, the fact is verified by the equation
$F_7 + F_4 = 13 + 3 = 16 = 2(8) = 2F_6$.

(b) Denoting the first of these four Fibonacci numbers by F_N, a mathematical formula expressing this fact is $F_{N+3} + F_N = 2F_{N+2}$. (The list in this particular case would be given by F_N, F_{N+1}, F_{N+2}, and F_{N+3}.)

17. (a) $2F_6 - F_7 = 2 \times 8 - 13 = 3 = F_4$

(b) $2F_{N+2} - F_{N+3} = F_N$
(The list in this particular case would be given by F_N, F_{N+1}, F_{N+2}, and F_{N+3}.)

B. The Golden Ratio

19. (a) $21\left(\frac{1+\sqrt{5}}{2}\right)+13 \approx 46.97871$

The order of operations is critical. First add 1 and $\sqrt{5}$. Next, divide by 2. Then multiply by 21. Finally, add 13.

(b) $\left(\frac{1+\sqrt{5}}{2}\right)^8 \approx 46.97871$

Order of operations is again critical. First add 1 and $\sqrt{5}$. Next, divide by 2. Finally, take the result to the power of 8.

(c) $\frac{\left(\frac{1+\sqrt{5}}{2}\right)^8-\left(\frac{1-\sqrt{5}}{2}\right)^8}{\sqrt{5}} = 21.00000$

21. (a) $\frac{\phi^8}{\sqrt{5}} = \frac{\left[\frac{(1+\sqrt{5})}{2}\right]^8}{\sqrt{5}} \approx 21$

Order of operations is critical. First add 1 and $\sqrt{5}$. Next, divide by 2. Next, take the result to the power of 8. Finally, divide the result by $\sqrt{5}$.

(b) $\frac{\phi^9}{\sqrt{5}} = \frac{\left[\frac{(1+\sqrt{5})}{2}\right]^9}{\sqrt{5}} \approx 34$

(c) Guess the 7th Fibonacci number, 13.

23. (a)
$$\begin{aligned}\phi^6 &= \phi^5 \cdot \phi\\ &= (5\phi+3)\phi\\ &= 5\phi^2+3\phi\\ &= 5(\phi+1)+3\phi\\ &= 5\phi+5+3\phi\\ &= 8\phi+5\end{aligned}$$

(b)
$$\begin{aligned}\phi^6 &= 8\phi+5\\ &= 8\left(\frac{1+\sqrt{5}}{2}\right)+5\\ &= 4+4\sqrt{5}+5\\ &= 4\sqrt{5}+9\end{aligned}$$

So, $a = 4$, $b = 9$.

25.
$$\begin{aligned}F_{500} &\approx \phi \times F_{499}\\ &\approx 1.6180 \times 8.6168 \times 10^{103}\\ &\approx 13.94 \times 10^{103}\\ &\approx 1.394 \times 10^{104}\end{aligned}$$

Remember that when using scientific notation, only one digit may occur before the decimal point.

27. (a)
$$\begin{aligned}A_7 &= 2A_{7-1}+A_{7-2}\\ &= 2A_6+A_5\\ &= 2(70)+29\\ &= 169\end{aligned}$$

(b) $\frac{A_7}{A_6} = \frac{169}{70}$
≈ 2.41429

(c) $\frac{A_{11}}{A_{10}} = \frac{5,741}{2,378}$
≈ 2.41421

(d) 2.41421 (Though not obvious, the ratio actually approaches $1+\sqrt{2}$ as the value of N increases.)

C. Fibonacci Numbers and Quadratic Equations

29. (a)
$$\begin{aligned}x^2 &= 2x+1\\ x^2-2x-1 &= 0\\ x &= \frac{-(-2)\pm\sqrt{(-2)^2-4(1)(-1)}}{2(1)}\\ &= \frac{2\pm\sqrt{8}}{2}\\ &= \frac{2\pm2\sqrt{2}}{2}\\ &= 1\pm\sqrt{2}\end{aligned}$$

$x = 1+\sqrt{2} \approx 2.41421$ or $x = 1-\sqrt{2} \approx -0.41421$

(b) $5x^2 = 2x + 1$

$5x^2 - 2x - 1 = 0$

$$x = \frac{-(-2) \pm \sqrt{(-2)^2 - 4(5)(-1)}}{2(5)}$$

$$= \frac{2 \pm \sqrt{24}}{10}$$

$$= \frac{1 \pm \sqrt{6}}{5}$$

So, $x = \frac{1+\sqrt{6}}{5} \approx 0.68990$,

$x = \frac{1-\sqrt{6}}{5} \approx -0.28990$

(c) $3x^2 = 8x + 5$

$3x^2 - 8x - 5 = 0$

$$x = \frac{-(-8) \pm \sqrt{(-8)^2 - 4(3)(-5)}}{2(3)}$$

$$= \frac{8 \pm \sqrt{124}}{6}$$

$$= \frac{4 \pm \sqrt{31}}{3}$$

So, $x = \frac{4+\sqrt{31}}{3} \approx 3.18925$,

$x = \frac{4-\sqrt{31}}{3} \approx$ -0.52259

31. (a) $55(1)^2 = 34(1) + 21$

$55 = 55$

$x = 1$ is a solution.

(b) Rewrite the equation:

$55x^2 - 34x - 21 = 0$

Then

$$1 + x = \frac{34}{55}$$

$$x = \frac{34}{55} - 1$$

$$= -\frac{21}{55}$$

$$\approx -0.38182$$

33. (a) Putting $x = 1$ in the equation gives $F_N = F_{N-1} + F_{N-2}$ which is a defining equation for the Fibonacci numbers.

(b) Rewrite the equation:

$F_N x^2 - F_{N-1} x - F_{N-2} = 0$

Then, using the hint in Exercise 31(b),

$$1 + x = \frac{F_{N-1}}{F_N}$$

$$x = \left(\frac{F_{N-1}}{F_N}\right) - 1$$

D. Similarity

35. (a) Since R and R' are similar, each side length of R' is 3 times longer than the corresponding side in R. So, the perimeter of R' will be 3 times greater than the perimeter of R. This means that the perimeter of R' is $3 \times 41.5 = 124.5$ inches.

(b) Since each side of R' is 3 times longer than the corresponding side in R, the area of R' will be $3^2 = 9$ times larger than the area of R. That is, the area of R' is $9 \times 105 = 945$ square inches.

37. (a) The ratio of the perimeters must be the same as the ratio of corresponding side lengths.

$$\frac{P}{13 \text{ in.}} = \frac{60 \text{ m}}{5 \text{ in.}}$$

$$P = \frac{60 \text{ m}}{5 \text{ in.}} \times 13 \text{ in.}$$

$$= 156 \text{ m}$$

(b) We use the fact that the ratio of the areas of two similar triangles (or any two similar polygons) is the same as the ratio of their side lengths squared. So,

$$\frac{A}{20 \text{ sq. in.}} = \left(\frac{60 \text{ m}}{5 \text{ in.}}\right)^2$$

$$A = \left(\frac{60 \text{ m}}{5 \text{ in.}}\right)^2 \times (20 \text{ sq. in.})$$

$$= 2880 \text{ sq. m}$$

39. There are two possible cases that need to be considered. First, we solve

$$\frac{3}{x} = \frac{5}{8-x}$$

$$24 - 3x = 5x$$

$$24 = 8x$$

$$3 = x$$

But, it could also be the case that the side of length 3 does not correspond to the side of length 5, but rather the side of length 8 – x. In this case, we solve

$$\frac{3}{x} = \frac{8-x}{5}$$
$$15 = 8x - x^2$$
$$x^2 - 8x + 15 = 0$$
$$(x-5)(x-3) = 0$$

So, $x = 3,5$. It follows that R and R' are similar if $x = 3$ or if $x = 5$ (as can easily be checked).

E. Gnomons

41. $$\frac{3}{9} = \frac{9}{c+3}$$
$$3(c+3) = 81$$
$$3c + 9 = 81$$
$$3c = 72$$
$$c = 24$$

43. $$\frac{8}{12} = \frac{3+8+1}{2+12+x}$$
$$\frac{8}{12} = \frac{12}{14+x}$$
$$8(14+x) = 144$$
$$112 + 8x = 144$$
$$8x = 32$$
$$x = 4$$

45.

A B
20
10 x
10 + x

$$\frac{10}{20} = \frac{20}{10+x}$$
$$10(10+x) = 400$$
$$100 + 10x = 400$$
$$10x = 300$$

So, $x = 30$. Rectangle B is 20 by 30.

47. **(a)** In the figure, the measure of angle CAD must be $180^\circ - 108^\circ = 72^\circ$. Since triangle BDC must be similar to triangle BCA (in order for triangle ACD to be a gnomon), it must be that the measure of angle BDC must be 36°. Using the fact that the sum of the measures of the angles in any triangle is 180°, it follows that the measure of angle ACD is 72°.

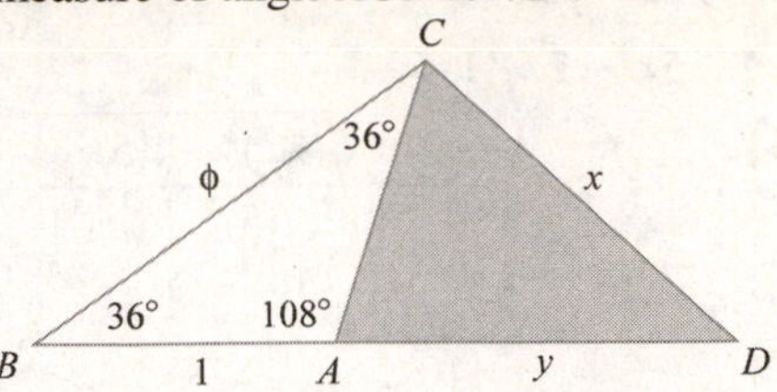

(b) Since triangle DBC is isosceles, $x = \phi$. Note that triangle ACD is also isosceles. Hence, $y = x = \phi$ as well. Likewise, triangle ABC is isosceles so that the length of AC is 1.

49. $\frac{3}{4} = \frac{9}{x}$ and $\frac{3}{5} = \frac{9}{5+y}$

$3x = 36$ and $3(5+y) = 45$

$x = 12$ and $y = 10$

JOGGING

51. $A_N = 5F_N$ (Each term in this sequence is 5 times more than the corresponding term in the Fibonacci sequence.)

53. **(a)**
$$T_1 = 7F_2 + 4F_1 = 7\cdot1 + 4\cdot1 = 11$$
$$T_2 = 7F_3 + 4F_2 = 7\cdot2 + 4\cdot1 = 18$$
$$T_3 = 7F_4 + 4F_3 = 7\cdot3 + 4\cdot2 = 29$$
$$T_4 = 7F_5 + 4F_4 = 7\cdot5 + 4\cdot3 = 47$$
$$T_5 = 7F_6 + 4F_5 = 7\cdot8 + 4\cdot5 = 76$$
$$T_6 = 7F_7 + 4F_6 = 7\cdot13 + 4\cdot8 = 123$$
$$T_7 = 7F_8 + 4F_7 = 7\cdot21 + 4\cdot13 = 199$$
$$T_8 = 7F_9 + 4F_8 = 7\cdot34 + 4\cdot21 = 322$$

(b)
$$T_N = 7F_{N+1} + 4F_N$$
$$= 7(F_N + F_{N-1}) + 4(F_{N-1} + F_{N-2})$$
$$= (7F_N + 4F_{N-1}) + (7F_{N-1} + 4F_{N-2})$$
$$= T_{N-1} + T_{N-2}$$

(c) $T_1 = 11$, $T_2 = 18$,
$T_N = T_{N-1} + T_{N-2}$

55. $$\frac{1+\sqrt{5}}{2} + \frac{1-\sqrt{5}}{2} = \frac{1+1}{2} = 1$$

Since the first term is positive and the second term is negative and the sum is a whole number, it follows that the decimal parts are the same.

57. We use the fact that $\phi^N = F_N\phi + F_{N-1}$ to compute the ratio *l/s*. Since $\frac{l}{s} = \frac{144\phi+89}{89\phi+55} = \frac{\phi^{12}}{\phi^{11}} = \phi$, it follows that *R* is a golden rectangle.

59. If the rectangle is a gnomon to itself, then the rectangle in the figure below must be similar to the original *l* by *s* rectangle.

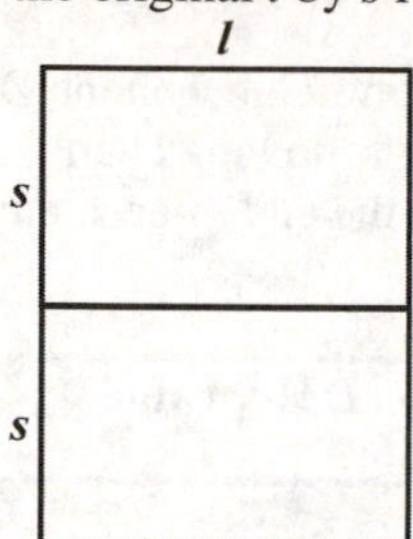

This means that $\frac{2s}{l} = \frac{l}{s}$. Letting $x = \frac{l}{s}$, we solve $\frac{2}{x} = x$. The only positive solution to this equation is $x = \frac{l}{s} = \sqrt{2}$. So, the only rectangles that are gnomons to themselves (homognomic) are those having this ratio.

61. Since the area of the white triangle is 6, the area of the shaded figure must be 48, which makes the area of the new larger similar triangle 54. Since the ratio of the areas of similar triangles is the square of the ratio of the sides, we have

$\frac{3+x}{3} = \sqrt{\frac{54}{6}} = 3$ and $\frac{y}{4} = \sqrt{\frac{54}{6}} = 3$ and

$\frac{5+z}{5} = \sqrt{\frac{54}{6}} = 3$ so $x = 6$, $y = 12$, $z = 10$.

63. Since the area of the white rectangle is 60 and the area of the shaded figure is 75, the area of the new larger similar rectangle is 135. Since the ratio of the area of similar rectangles is the square of the ratio of the sides, we have

$\frac{6+x}{6} = \sqrt{\frac{135}{60}} = \sqrt{\frac{9}{4}} = \frac{3}{2}$ and $\frac{10+y}{10} = \frac{3}{2}$ so $x = 3$, $y = 5$.

65. **(a)** Since we are given that $AB = BC = 1$, we know that $\angle BAC = 72°$ and so $\angle BAD = 180° - 72° = 108°$. This makes $\angle ABD = 180° - 108° - 36° = 36°$ and so triangle *ABD* is isosceles with $AD = AB = 1$. Therefore $AC = x - 1$. Using these facts and the similarity of triangle *ABC* and triangle *BCD* we have $\frac{x}{1} = \frac{1}{x-1}$ or, $x^2 = x + 1$ for which we know the solution is $x = \phi$.

(b) 36°, 36°, 108°

(c) $\frac{\text{longer side}}{\text{shorter side}} = \frac{x}{1} = x = \phi$

67. **(a)** The regular decagon can be split into ten $72° - 72° - 36°$ triangles as shown in the figure. In each of these triangles, the side opposite the 72° angle is the same length as the radius of the circle, namely *r*. In exercise 65, it was shown that the length of the side opposite the 36° angle is $\frac{r}{\phi}$. Hence, the perimeter of the regular decagon is $10 \times \frac{1}{\phi} = \frac{10}{\phi}$.

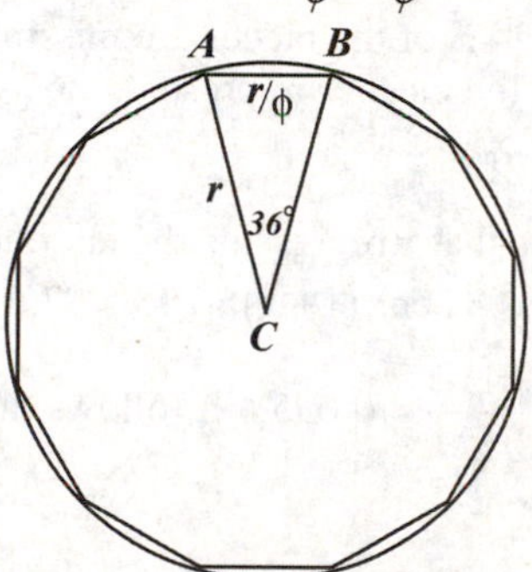

(b) Using part (a), the perimeter is found to be $\frac{10r}{\phi}$.

69. The sum of two odd numbers is always even and the sum of an odd number and an even number is always odd. Since the seeds in the Fibonacci sequence are both odd, every third number is even while the others are all odd.

Chapter 10

WALKING

A. Percentages

1. (a) 0.0225 (drop the % sign and move the decimal point two places to the left)

(b) 0.0075

3. (a) This is the same as 2.25 out of every 100 or 225 out of every 10,000. So,

$$2.25\% = \frac{2.25}{100} = \frac{225}{10,000} = \frac{9}{400}.$$

(b) $0.75\% = \frac{75}{10,000} = \frac{3}{400}$

5. 80%; $\frac{60}{75} = \frac{x}{100}$ gives $x = 80$.

7. 215; If 14% of the pieces are missing, then 86% of the pieces are present. 86% of 250 is $0.86 \times 250 = 215$.

9. Suppose that x represents the tax rate on the earrings. Then, $(1+x)\$6.95 = \$7.61$. So, $x = \frac{\$7.61}{\$6.95} - 1 = 0.095$. It follows that the tax rate is 9.5%.

11. \$12.5 trillion; $\frac{16\%}{100\%} = \frac{\$2,000,000,000,000}{x}$ gives $x =$ \$12,500,000,000,000.

13. If T was the starting tuition, then the tuition at the end of one year is 110% of T. That is, $(1.10)T$. The tuition at the end of two years is 115% of what it was after one year. That is, after two years, the tuition was $(1.15)(1.10)T$. Likewise, the tuition at the end of the three years was
$(1.10)(1.15)(1.10)T = 1.3915T$.
The total percentage increase was 39.15%.

15. Suppose that the cost of the shoes is originally \$$C$. After the mark up, the price is \$$2.2C$. When the shoes go on sale, they are priced at 80% of that price or \$$(0.8)(2.2)C$. When the shoes find their way to the clearance rack, they are priced at 70% of the sale price which is \$$(0.7)(0.8)(2.2)C$. After the customer applies the coupon, the final price is 90% of the clearance price. That is,
\$$(0.9)(0.7)(0.8)(2.2)C$ = \$$1.1088C$.
So, the profit on these shoes is 10.88% more than the original cost \$$C$. [If you still don't believe it, imagine the shoes are priced at an even \$100 originally and do the computations.]

17. Suppose that when the week began the DJIA had a value of A. The following chart describes the value at the end of each day of the week.

Day	DJIA value
Mon.	$(1.025)A$
Tues.	$(1.121)(1.025)A$
Wed.	$(0.953)(1.121)(1.025)A$
Thur.	$(1.008)(0.953)(1.121)(1.025)A$
Fri.	$(0.946)(1.008)(0.953)(1.121)(1.025)A$

At the end of the week the DJIA has a value of approximately $1.044A$, an net increase of 4.4%.

19. (a) \$46,000,000;
$\frac{17.7\%}{100\%} = \frac{\$8,142,000}{x}$ gives $x =$ \$46,000,000.

(b) Because he made more than \$94,200 in income, Buffet (only) paid $0.062 \times \$94,200 = \$5840.40$ in social security tax in 2006. With an income of \$46,000,000 (see part (a)), this is roughly $\frac{\$5840.40}{\$46,000,000} = 0.000127$ or 0.0127% of his income.

B. Simple Interest

21. (a) $\$875(1+0.0428\times 4)$ = \$1,024.80

(b) $\$875(1+0.0428\times 5)$ = \$1,062.25

(c) $\$875(1+0.0428\times 10)$ = \$1,249.50

23. When you cash in the bond after five years, it is worth $\$10,000(1+0.05\times 5) = \$12,500$.

After paying 15% in federal taxes on the \$2,500 of interest earned, you keep 85% of the interest or $0.85\times \$2,500 = \$2,125$. This means that you net \$12,125 (\$2,125 more than you paid for the bond).

25. We solve $\$P(1+0.0575\times 4) = \$6000$ for P to find $\$P = \dfrac{\$6000}{(1+0.0575\times 4)} = \$4,878.05$.

27. Using the Simple Interest Formula, we solve $\$5400(1+r\cdot 8) = \$8316$ for r to find $r = \dfrac{\$8316\text{-}\$5400}{\$5400\times 8} = 0.0675$. So, the APR is 6.75%.

29. If the principal doubles in 12 years, then $F = 2P$ can be used in the Simple Interest Formula. We solve $\$P(1+r\cdot 12) = \$2P$ for r to find $r = \dfrac{\$2P\text{-}\$P}{\$P\times 12} = \dfrac{1}{12} \approx 0.0833$. This gives an APR of 8.33% (to the nearest hundredth percent).

C. Compound Interest

31. (a) $\$3250\times(1+0.09)^4 = \$4587.64$

(b) Since interest is credited to the account at the end of each year, no growth takes place during the last seven months. The balance after this time is thus $\$3250\times(1+0.09)^5 = \$5000.53$.

33. The initial deposit of \$3420 grows at a $6\frac{5}{8}$% annual interest rate for 3 years. At the end of three years (on Jan. 1, 2011), the amount in the account is $\$3420\times(1+0.06625)^3 = \$4145.75$. This amount then grows at a $5\frac{3}{4}$% annual interest rate for 4 years. At the end of that time (on Jan. 1, 2015), the amount in the account is $\$4145.75\times(1.0575)^4 = \$5184.71$.

35. The balance on December 31, 2006 is $\$3420\times(1.06625)^2 = \$3888.16$. The balance on January 1, 2007 is $\$3888.16 - \$1500 = \$2388.16$.
The balance on December 31, 2007 is $\$2388.16\times(1.06625)^1 = \$2546.38$.
The balance on January 1, 2008 is $\$2546.38 - \$1000 = \$1546.38$.
The balance on January 1, 2011 is $\$1546.38\times(1.06625)^3 = \$1874.53$.

37. (a) The periodic interest rate (here $r = 0.12$ and $n = 12$) is $p = \dfrac{r}{n} = \dfrac{0.12}{12} = 0.01$ (i.e. 1%). The total number of times T the principal is compounded over the life of the investment is $T = n\cdot t = 12\times 5 = 60$. Using the General Compounding Formula (version 2) with $P = \$5000$, $T = 60$, and $p = 0.01$, the future value of the account after 5 years is
$\$5000\times(1+0.01)^{60} = \$9083.48$

(b) The future value of \$1 in 1 year (12 months of growth at 1% interest each month) is $\$(1+0.01)^{12} \approx \$1.126825$
So, the APY is approximately 12.6825%.

39. (a) First, the number of hours in one year is
$1\text{ year}\times\dfrac{365\text{ days}}{1\text{ year}}\times\dfrac{24\text{ hours}}{1\text{ day}} = 8760\text{ hours}$
The General Compounding Formula gives a future value of
$\$875\times\left(1+\dfrac{0.0675}{8760}\right)^{7.5\times 8760} = \$1451.67$.

(b) The future value of \$1 in 1 year (8760 hours of growth) is
$\$\left(1+\dfrac{0.0675}{8760}\right)^{8760} \approx \$1.06983$
This is an APY of approximately 6.983%.

41. APY of option 1: 6%;
APY of option 2:
Each \$1 invested grows to
$\$\left(1+\dfrac{0.0575}{12}\right)^{12} \approx \$1.05904$
The APY is thus 5.904%;
APY of option 3:
Each \$1 invested grows to $\$e^{0.055} \approx \$1.05654$
The APY is thus 5.654%.

43. At an APR of 12%, the APYs are as follows:

Compounding	APY
Yearly	12%
Semiannually	12.36%
Quarterly	12.5509%
Monthly	12.6825%
Daily	12.7475%
Hourly	12.7496%
Continuously	12.7497%

45. $\frac{\$810-\$750}{\$750}=0.08$ so the APY is 8%.

47. (a) We use the General Compounding Formula to find the value of t when F = \$2000, P = \$1000, and r = 0.06. Trial and error gives
$\$2000=\$1000(1+0.06)^t$ when $t=12$ years.

(b) This is the same as (a) only we use $F = 2P$ since P is unknown. Solving $2P=P(1+0.06)^t$ for t (again, by trial and error) gives the same answer of 12 years.

49. We solve $\$4060.17=P(1+0.0475)^3$ for P.

This gives $P=\frac{\$4060.17}{(1+0.0475)^3}=\$3532.50$.

D. Geometric Sequences and The Geometric Sum Formula

51. (a) $G_1=11\times1.25=13.75$

(b) $G_6=11\times(1.25)^6\approx41.962$

(c) $G_N=11\times(1.25)^N$

53. (a) $c=\frac{G_1}{G_0}=\frac{2304}{3072}=\frac{3}{4}$

(b) $G_4=3072\times\left(\frac{3}{4}\right)^4=972$

(c) $G_N=3072\times\left(\frac{3}{4}\right)^N$

55. (a) $G_N=(1.50)\cdot G_{N-1}$ where $G_0=256$ represents the number of crimes committed in 2000 and G_N represents the number of crimes committed in the year 2000 + N.
We multiply by 1.50 each year to account for an *increase* by 50%. To determine how many crimes were committed in 2005, we compute

$$\begin{aligned}G_5&=(1.50)^5\cdot G_0\\&=(1.50)^5\times256\\&=1944\end{aligned}$$

So, 1944 crimes were committed in 2005.

(b) $G_N=(1.5)^N\cdot256$

(c) $G_{10}=(1.50)^{10}\times256\approx14{,}762$. So, approximately 14,762 crimes will be committed in 2010.

57. (a) $G_{100}=3\times2^{100}$

(b) $G_N=3\times2^N$

(c) Use P = 3, c = 2, and N = 101 in the Geometric Sum Formula.

$$\begin{aligned}&G_0+G_1+\ldots+G_{100}\\&=P+c\cdot P+c^2\cdot P+\cdots+c^{100}\cdot P\\&=3+2\times3+2^2\times3+\ldots+2^{100}\times3\\&=3\times\left(\frac{2^{101}-1}{2-1}\right)\\&=3\times(2^{101}-1)\end{aligned}$$

(d) Use $P=3\times2^{50}$, c = 2, and N = 51 in the Geometric Sum Formula.

$$\begin{aligned}&G_{50}+G_{51}+\ldots+G_{100}\\&=P+c\cdot P+c^2\cdot P+\cdots+c^{50}\cdot P\\&=3\times2^{50}+3\times2^{51}+3\times2^{52}+\ldots+3\times2^{100}\\&=(3\times2^{50})\times\left(\frac{2^{51}-1}{2-1}\right)\\&=3\times2^{50}\times(2^{51}-1)\end{aligned}$$

59. In this geometric sum, we have $P = \$10(1.05)$, $c = 1.05$, and $N = 36$. Thus,

$$\$10(1.05)+\$10(1.05)^2+\ldots+\$10(1.05)^{36}$$
$$=\$10(1.05)\frac{1.05^{36}-1}{1.05-1}$$
$$=\$1006.28$$

61. In this geometric sum, we have $P=\frac{10}{1.05}$, $c=\frac{1}{1.05}$, and $N = 36$. Thus,

$$\frac{\$10}{1.05}+\frac{\$10}{1.05^2}+\ldots+\frac{\$10}{1.05^{35}}+\frac{\$10}{1.05^{36}}$$
$$=\frac{\$10}{1.05}\left(\frac{\left(\frac{1}{1.05}\right)^{36}-1}{\frac{1}{1.05}-1}\right)$$
$$=\$165.47$$

E. Deferred Annuities

63. We use the Fixed Deferred Annuity Formula with $T = 40$ payments (deposits) of $\$P = \$2000$ having a periodic (yearly) interest rate of $p = 0.075$. Since payments are made at the beginning of each period, the future value of the last payment is

$$L=(1+p)P$$
$$=(1+0.075)\times\$2000$$
$$=\$2150$$

So, the future value of Markus' IRA when he retires at the age of 65 is

$$F=L\left[\frac{(1+p)^T-1}{p}\right]$$
$$=\$2150\left[\frac{(1.075)^{40}-1}{0.075}\right]$$
$$\approx\$488,601.52$$

65. Just as in exercise 63, we use the Fixed Deferred Annuity Formula with $T = 40$ payments (deposits) of $\$P = \$2000$ having a periodic (yearly) interest rate of $p = 0.075$. However, since payments are made at the *end* of each period, the future value of the last payment is $L = P = \$2000$.
So, the future value of Markus' IRA when he retires at the age of 65 is now

$$F=L\left[\frac{(1+p)^T-1}{p}\right]$$
$$=\$2000\left[\frac{(1.075)^{40}-1}{0.075}\right]$$
$$\approx\$454,513.04$$

67. We use the Fixed Deferred Annuity Formula with $T = 11$ payments (deposits) of $\$P = \$100$ having a periodic (monthly) interest rate of $p = 0.06/12 = 0.005$. Since payments are made at the beginning of each period, the future value of the last payment is

$$L=(1+p)P$$
$$=(1+0.005)\times\$100$$
$$=\$100.50$$

So, the future value of the Christmas Club account on December 1st is

$$F=L\left[\frac{(1+p)^T-1}{p}\right]$$
$$=\$100.50\left[\frac{(1.005)^{11}-1}{0.005}\right]$$
$$\approx\$1133.56$$

69. We use the Fixed Deferred Annuity Formula with $T = 36$ payments of $\$P = \$400$ having a monthly interest rate of $p = 0.045/12 = 0.00375$. Since payments are made at the end of each period, the future value of the last payment is $L = P = \$400$.
So, the future value of Celine's deposits at the end of three years is

$$F=L\left[\frac{(1+p)^T-1}{p}\right]$$
$$=\$400\left[\frac{(1.00375)^{36}-1}{0.00375}\right]$$
$$\approx\$15,386.44$$

If this is a 20% down payment, then the maximum price of a house she can buy is five times as much or $76,932.20.

71. In this exercise, we determine the value of $P in the Fixed Deferred Annuity Formula where $T=12\times52=624$ payments (deposits) having a periodic (weekly) interest rate of $p = 0.0468/52 = 0.0009$. Since payments are made at the beginning of each week, the future value of the last payment is $L=(1+p)P=1.0009P$.

The future value of the college trust fund after 12 years is (hopefully)

$$\$150,000 = L\left[\frac{(1+p)^T - 1}{p}\right]$$
$$= \$1.0009P\left[\frac{(1.0009)^{624} - 1}{0.0009}\right]$$

Solving this for P gives Layla's required weekly deposit:

$$\$P = \frac{\$150,000}{1.0009}\left[\frac{0.0009}{(1.0009)^{624} - 1}\right] = \$179.11$$

F. Installment Loans

73. **(a)** We use the Amortization Formula with $T = 25$ payments of $\$F = \$3000$ having a periodic (yearly) interest rate of $p = 0.06$. So, if $q = \frac{1}{1+p} = \frac{1}{1.06}$, then such an installment loan has a present value of

$$\$P = \$Fq\left[\frac{q^T - 1}{q - 1}\right]$$
$$= \$3000\left(\frac{1}{1.06}\right)\left[\frac{\left(\frac{1}{1.06}\right)^{25} - 1}{\left(\frac{1}{1.06}\right) - 1}\right]$$
$$\approx \$38,350.07$$

(b) This is similar to (a) only having an annual interest rate of $p = 0.07$. So, if $q = \frac{1}{1.07}$, then such an installment loan has a present value of

$$\$P = \$3000\left(\frac{1}{1.07}\right)\left[\frac{\left(\frac{1}{1.07}\right)^{25} - 1}{\left(\frac{1}{1.07}\right) - 1}\right]$$
$$\approx \$34,960.75$$

75. We use the Amortization Formula with $T = 60 \times 52 = 3120$ (weekly) payments of $\$F = \$50$ having a periodic (weekly) interest rate of $p = 0.052/52 = 0.001$. So, if $q = \frac{1}{1+p} = \frac{1}{1.001}$, then such an annuity has a present value of

$$\$P = \$Fq\left[\frac{q^T - 1}{q - 1}\right]$$
$$= \$50\left(\frac{1}{1.001}\right)\left[\frac{\left(\frac{1}{1.001}\right)^{3120} - 1}{\left(\frac{1}{1.001}\right) - 1}\right]$$
$$\approx \$47,788.70$$

77. **(a)** In this exercise, we determine the value of F in the Amortization Formula with $T = 30 \times 12 = 360$ payments and a monthly interest rate of $p = 0.0575/12$. With $q = \frac{1}{1 + \frac{0.0575}{12}}$, such a mortgage loan has a present value of \$160,000 when $\$160,000 = \$Fq\left[\frac{q^{360} - 1}{q - 1}\right]$. That is,

when $$\$F = \frac{\$160,000}{q}\left[\frac{q - 1}{q^{360} - 1}\right]$$
$$\approx \$933.72$$

(b) The Simpsons will make 360 payments of \$933.72 for a total of \$336,139.20. This means they will pay \$336,139.20 - \$160,000 = \$176,139.20 in interest.

79. We find the present value of a fixed immediate annuity (i.e. an installment loan) and then add \$25,000. In this case, the payment is $F =$ \$950, $T = 12 \times 20 = 240$ months, and $p = 0.055/12$. So, if $q = \frac{1}{1 + \frac{0.055}{12}}$, then such an annuity has a present value of

$$\$P = \$Fq\left[\frac{q^T - 1}{q - 1}\right]$$
$$= \$950q\left[\frac{q^{240} - 1}{q - 1}\right]$$
$$\approx \$138,104$$

Adding the \$25,000 down payment, we find the selling price of the house that Ken just bought to be approximately \$163,104.

JOGGING

81. 100%

Let m be the markup and c be the retailer's

cost.

$$0.75(1+m)c = 1.5c$$
$$0.75 + 0.75m = 1.5$$
$$0.75m = 0.75$$
$$m = 1$$

The markup should be 100%.

83. **(a)** Another way to write $y\%$ is $y/100$ and another way to write $x\%$ is $x/100$. So, taking $y\%$ off an item is the same as multiplying the price of the item by 1-$y/100$. Taking $x\%$ off an item is the same as multiplying the price by 1-$x/100$. If the item originally cost \$$P$, then the result of these two discounts would be an item priced at $\$\left(1-\frac{x}{100}\right)\left[\left(1-\frac{y}{100}\right)P\right]$.

(b) $\$\left(1-\frac{y}{100}\right)\left[\left(1-\frac{x}{100}\right)P\right]$

(c) Multiplication is commutative. That is, it can be done in any order.

85. We use the Amortization Formula with $T = 10$ payments of $\$F = \$100{,}000{,}000$ having an annual interest rate of $p = 0.105$. So, if $q = \frac{1}{1+p} = \frac{1}{1.105}$, then such an installment loan has a present value of

$$\$P = \$Fq\left[\frac{q^T - 1}{q-1}\right]$$
$$= \$100{,}000{,}000\left(\frac{1}{1.105}\right)\left[\frac{\left(\frac{1}{1.105}\right)^{10} - 1}{\left(\frac{1}{1.105}\right) - 1}\right]$$
$$\approx \$601{,}477{,}274$$

87. As the periodic interest rate p goes up, more of the periodic payment must be used to pay interest. So, the value of the immediate annuity goes down. (From an algebraic point of view, if p goes up, then q goes down. In fact, though it is not easy to see, $\frac{q^T - 1}{q-1}$ also decreases as q decreases. So, the expression $\$Fq\left[\frac{q^T-1}{q-1}\right]$ will also decrease as it is a product of two expressions that each decrease as p increases.)

89. **(a)** This is the Geometric Sum Formula using $P(1+p)$ in place of P, $(1+p)$ in place of c, and T in place of N. As another way to see this (see exercise 88), multiply the left side by $(1 + p) - 1$ and the right side by p. You will see lots of cancellations on the left side.

(b) Move the location of the p in the denominator and simplify:

$$P(1+p)\left[\frac{(1+p)^T - 1}{p}\right] = P\left(\frac{1+p}{p}\right)\left[(1+p)^T - 1\right]$$
$$= P\left(\frac{1}{p} + 1\right)\left[(1+p)^T - 1\right]$$

Chapter 11

WALKING

A. Reflections

1. (a) 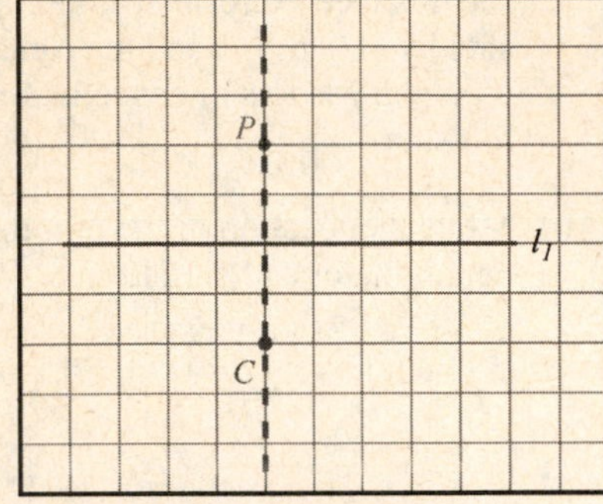

C. The image point of P under the reflection with axis l_1 is found by drawing a line through P perpendicular to l_1 and finding the point on this line on the opposite side of l_1 which is the same distance from l_1 as the point P. This point is C.

(b) F. See the figure below.

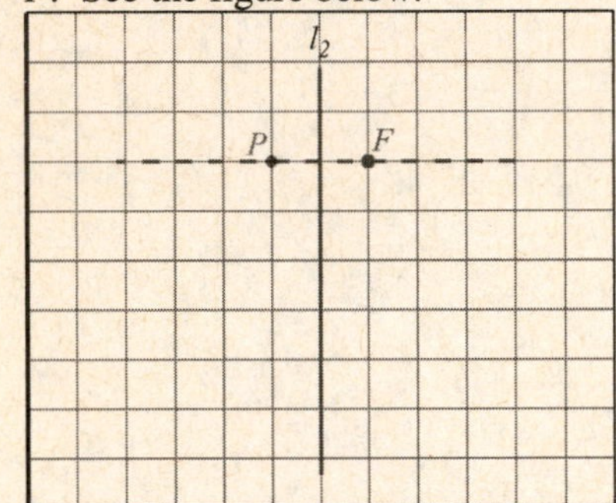

(c) E. See the figure below.

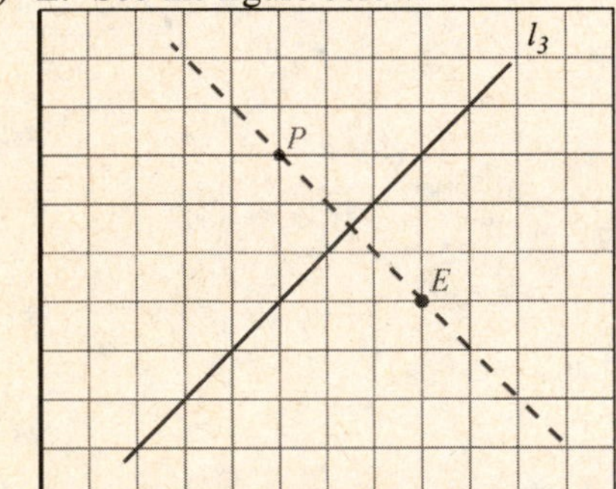

(d) B. See the figure below.

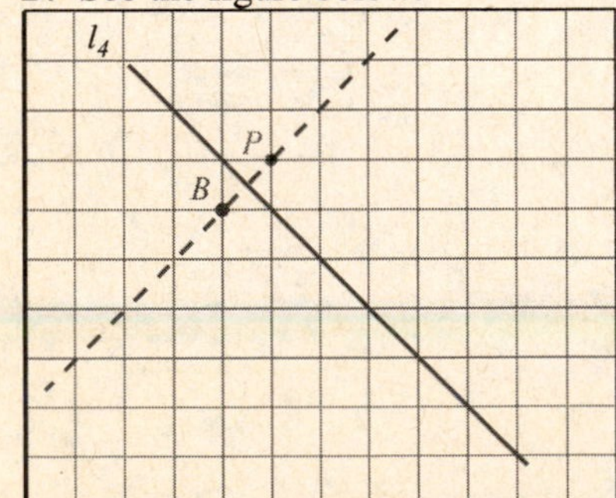

3. (a)
(b)
(c)

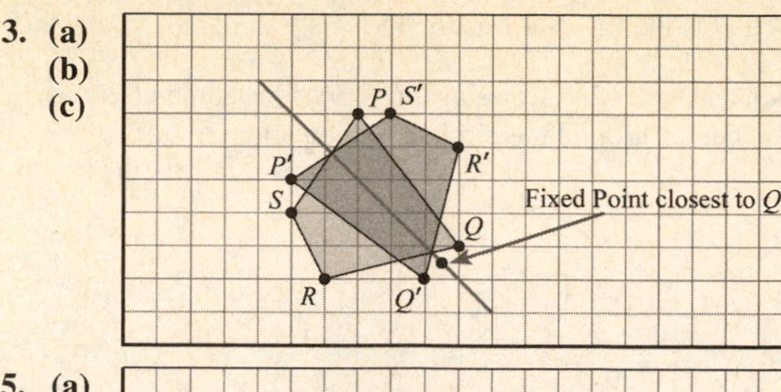

5. (a)
(b)
(c)
(d)

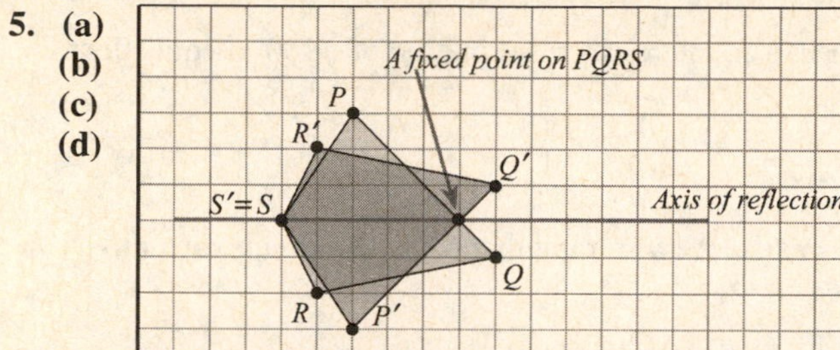

7. (a)
(b)

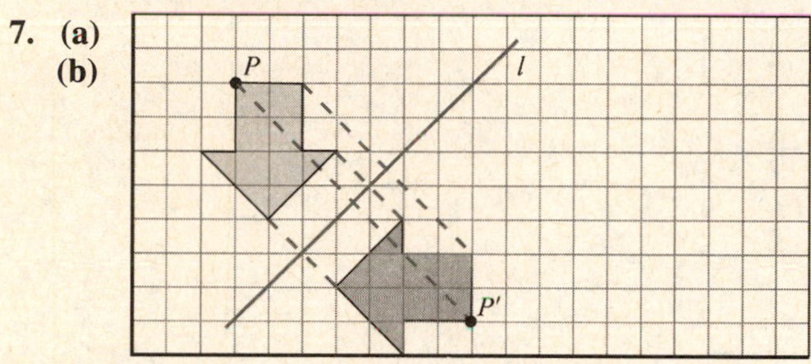

9.

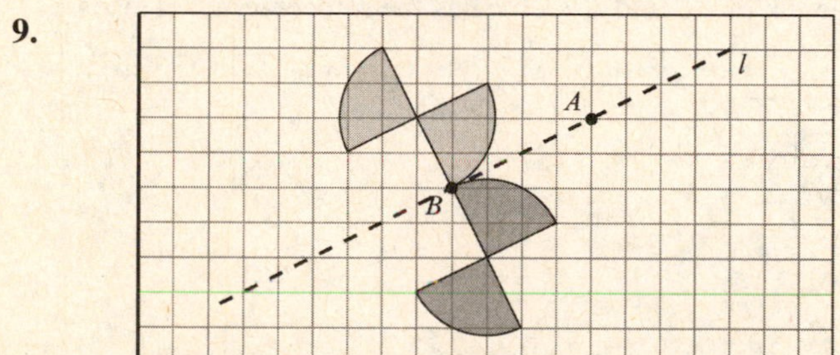

Since points *A* and *B* are fixed points, the axis of reflection *l* must pass through these points.

B. Rotations

11. The following figure may be of help in solving each part of this exercise.

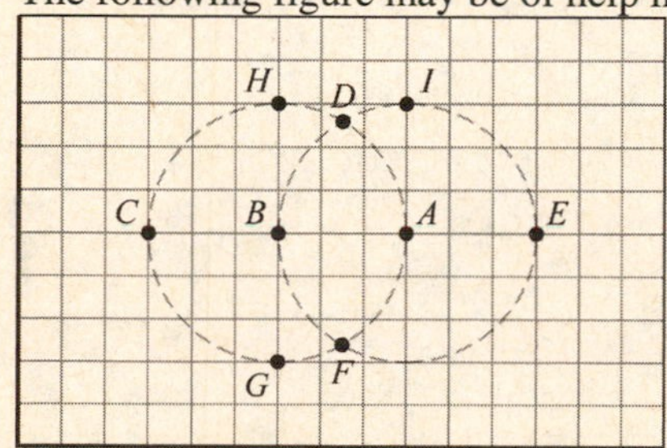

(a) *I*. Think of *A* as the center of a clock in which *B* is the "9" and *I* is the "12."

(b) *G*. Think of *B* as the center of a clock in which *A* is the "3" and *G* is the "6."

(c) *A*. Think of *B* as the center of a clock in which *D* is the "1" and *A* is the "3."

(d) *F*. Think of *B* as the center of a clock in which *D* is the "1" and *F* is the "5."

(e) *E*. Think of *A* as the center of a clock in which *I* is the "12." Rotating $3690°$ has the same effect as rotating $10\times360°+90°$. That is, as rotating 10 times around the circle and then another $90°$.

13. (a) $2(360°)-710°=10°$

(b) $710°-360°=350°$

(c) $\begin{array}{r}19\\ 360^\circ\overline{)7100^\circ}\end{array}$ with a remainder of 260°. Hence, a counterclockwise rotation of 7100° is equivalent to a clockwise rotation of $360^\circ - 260^\circ = 100^\circ$.

(d) $\begin{array}{r}197\\ 360^\circ\overline{)71{,}000^\circ}\end{array}$ with a remainder of 80°. Hence, a clockwise rotation of $71{,}000^\circ$ is equivalent to a clockwise rotation of 80°.

15. (a)
(b)

Since *BB'* and *CC'* are parallel, the intersection of *BC* and *B'C'* locates the rotocenter *O*. This is a 90° clockwise rotation.

17. (a)
(b)
(c)

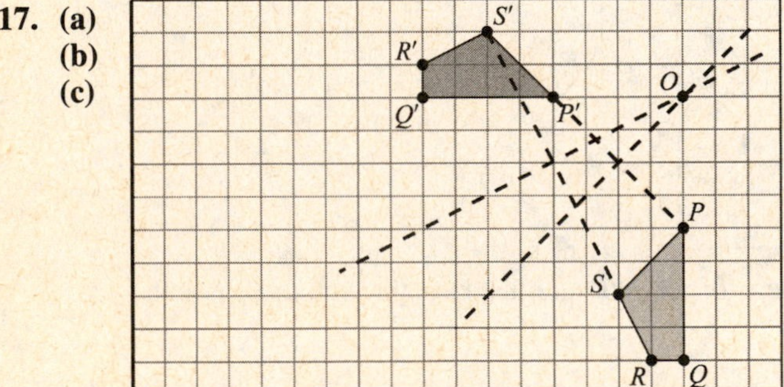

The rotocenter *O* is located at the intersection of the perpendicular bisectors to *PP'* and *SS'*. This is a 90° counterclockwise rotation.

19.

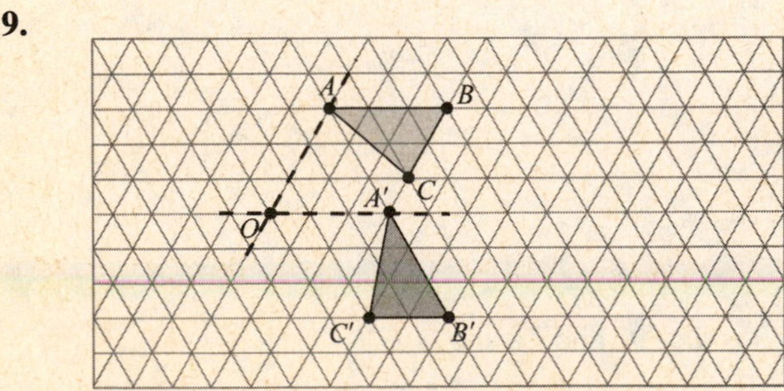

The equilateral triangles that make up the grid have interior angles that each measure 60°.

C. Translations

21. **(a)** *C*. Vector v_1 translates a point 4 units to the right so the image of *P* is *C*.

(b) *C*. Vector v_2 translates a point 4 units to the right so the image of *P* is *C*.

(c) *A*. Vector v_3 translates a point up 2 units and right 1 unit so the image of *P* is *A*.

(d) *D*. Vector v_4 translates a point down 2 units and left 1 unit so the image of *P* is *D*.

23.

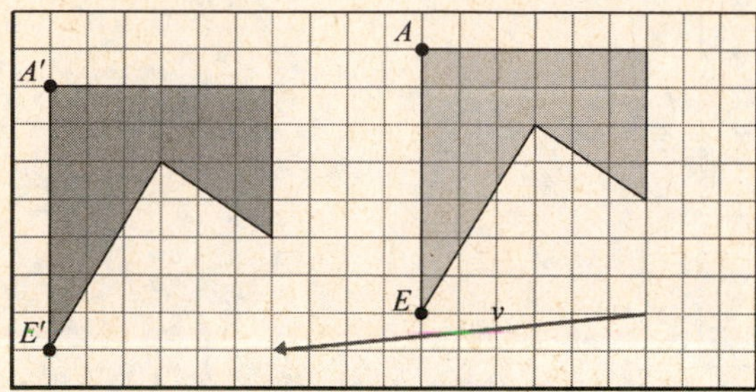

25.

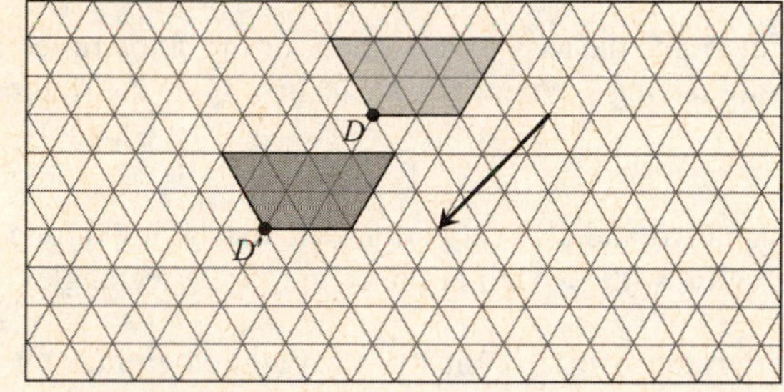

D. Glide Reflections

27.

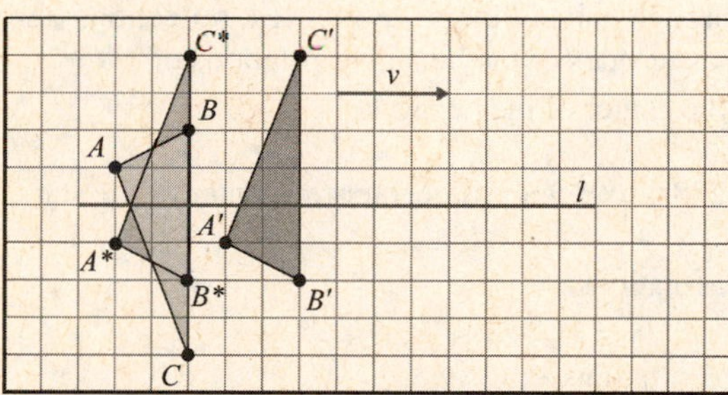

First, reflect the triangle *ABC* about the axis *l* (to form triangle *A*B*C**). Then, glide the figure three units to the right.

29. **(a)** **(b)** **(c)**

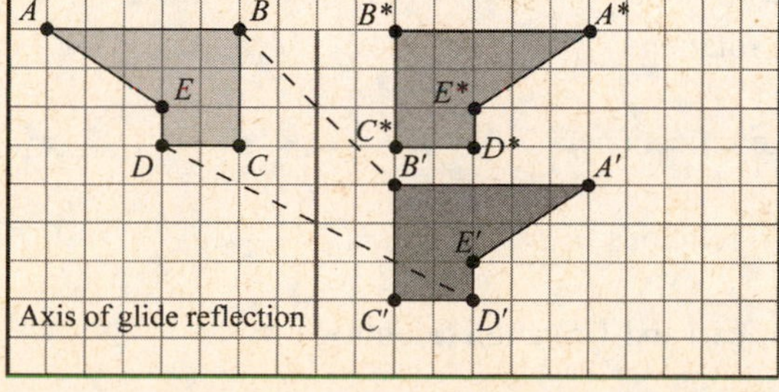

The midpoints of line segments *BB'* and *DD'* determine the axis of reflection for this glide reflection. First, reflect the figure *ABCDE* about the axis *l* (to form *A*B*C*D*E**). Then, glide the figure four units down.

31. (a)
(b)

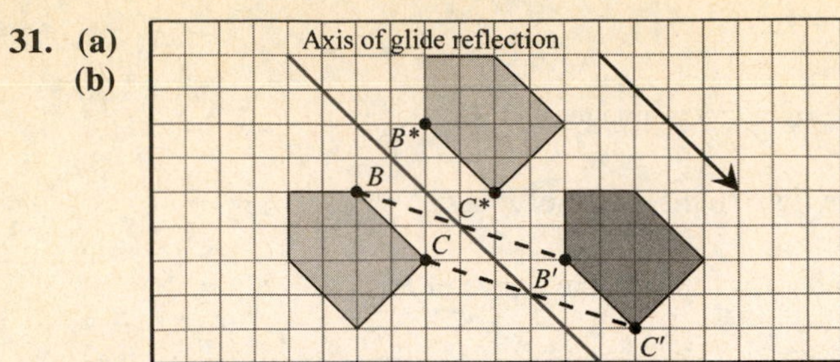

The midpoints of line segments *BB'* and *CC'* determine the axis of reflection for this glide reflection. First, reflect the figure about the axis. Then, glide (translate) the figure four diagonal units down and to the right (i.e. four units down and four units right).

33.

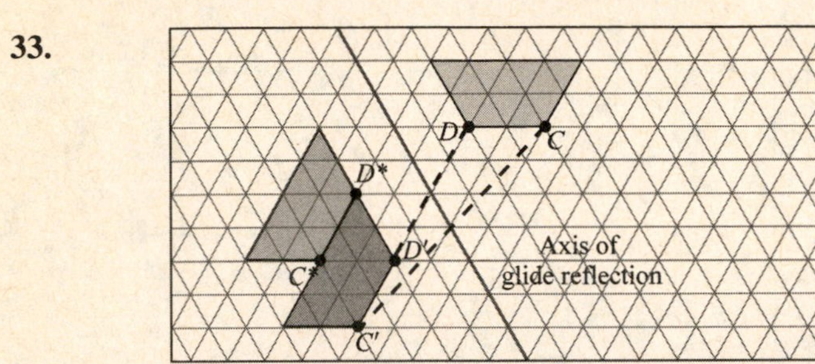

The midpoints of line segments *CC'* and *DD'* determine the axis of reflection for this glide reflection. First, reflect the figure about the axis (to the figure having segment *C*D**). Then, glide the figure.

E. Symmetries of Finite Shapes

35. (a) Reflection with axis going through the midpoints of *AB* and *DC*; reflection with axis going through the midpoints of *AD* and *BC*; rotations of 180° and 360° with rotocenter the center of the rectangle.

(b) No Reflections. Rotations of 180° and 360° with rotocenter the center of the parallelogram.

(c) Reflection with axis going through the midpoints of *AB* and *DC*; rotation of 360° with rotocenter the center of the trapezoid.

37. (a) Reflections (three of them) with axis going through pairs of opposite vertices; reflections (three of them) with axis going through the midpoints of opposite sides of the hexagon; rotations of 60°, 120°, 180°, 240°, 300°, 360° with rotocenter the center of the hexagon.

(b) No reflections; rotations of 72°, 144°, 216°, 288°, 360° with rotocenter the center of the star.

39. (a) D_2; the figure has exactly 2 reflections and 2 rotations.

(b) Z_2; the figure has no reflections and exactly 2 rotations.

(c) D_1; the figure has exactly 1 reflection and 1 rotation.

41. (a) Z_5; the figure has exactly 5 rotations (and no reflections).

(b) D_6; the figure has exactly 6 reflections and 6 rotations.

43. (a) D_1; the letter **A** has exactly 1 reflection (vertical) and 1 rotation (identity).

(b) D_1; the letter **D** has exactly 1 reflection (horizontal) and 1 rotation (identity).

(c) Z_1; the letter **L** has no reflection and exactly 1 rotation (identity).

(d) Z_2; the letter **Z** has no reflection and exactly 2 rotations (identity and $180°$).

(e) D_2; the letter **H** has exactly 2 reflections and 2 rotations (identity and $180°$).

(f) Z_2; the letter **N** has exactly no reflection and 2 rotations (identity and $180°$).

45. Answers will vary.

(a) Since symmetry type Z_1 has no reflections and exactly 1 rotation, the capital letter **J** is an example of this symmetry type.

(b) Since symmetry type D_1 has exactly 1 reflection and 1 rotation, the capital letter **T** is an example of this symmetry type.

(c) Since symmetry type Z_2 has no reflection and exactly 2 rotations, the capital letter **Z** is an example of this symmetry type.

(d) Since symmetry type D_2 has exactly 2 reflections and 2 rotations, the capital letter **I** is an example of this symmetry type.

47. Answers will vary.

(a) Symmetry type D_5 is common among many types of flowers (daisies, geraniums, etc.). The only requirements are that the flower have 5 equal, evenly spaced petals and that the petals have a reflection symmetry along their long axis. In the animal world, symmetry type D_5 is less common, but it can be found among certain types of starfish, sand dollars, and in some single celled organisms called diatoms.

(b) The Chrysler Corporation logo is a classic example of a shape with symmetry D_5. Symmetry type D_5 is also common in automobile wheels and hubcaps. One of the largest and most unusual buildings in Washington, DC has symmetry of type D_5.

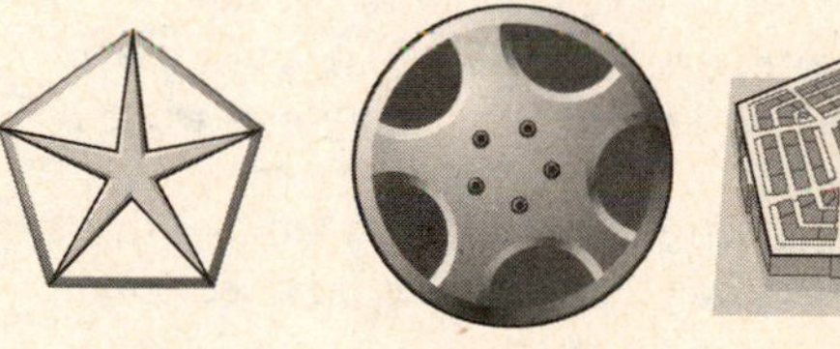

(c) Objects with symmetry type Z_1 are those whose only symmetry is the identity. Thus, any "irregular" shape fits the bill. Tree leaves, seashells, plants, and rocks more often than not have symmetry type Z_1.

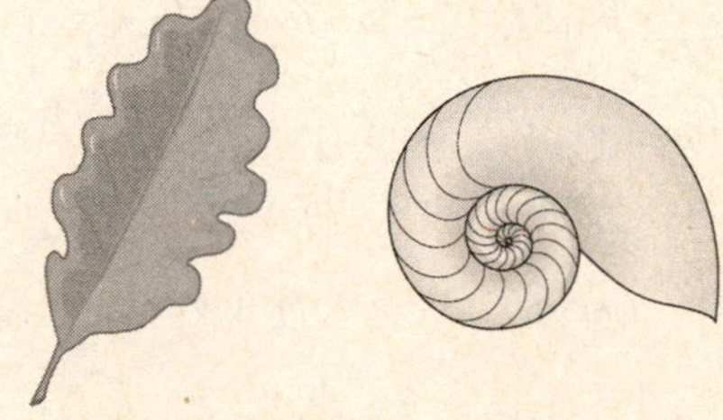

(d) Examples of manmade objects with symmetry of type Z_1 abound.

F. Symmetries of Border Patterns

49. **(a)** *m*1; the border pattern has translation symmetry and vertical reflection but does not have horizontal reflection, half-turn rotation, or glide reflection.

(b) 1*m*; the border pattern has translation symmetry and horizontal reflection but does not have vertical reflection, half-turn, or glide reflection.

(c) 12; the border pattern has translation symmetry and half-turn rotation but does not have horizontal reflection, vertical reflection, or glide reflection.

(d) 11; the border pattern has translation symmetry but does not have horizontal reflection, vertical reflection, half-turn rotation, or glide reflection.

51. **(a)** *m*1; the border pattern has translation symmetry and vertical reflection but does not have horizontal reflection, glide reflection, or half-turn rotation.

(b) 12; the border pattern has translation symmetry and half-turn rotation but does not have horizontal reflection, vertical reflection, or glide reflection.

(c) 1*g*; the border pattern has translation symmetry and glide reflection but does not have horizontal reflection, vertical reflection, or half-turn rotation.

(d) *mg*; the border pattern has translation symmetry, vertical reflection, half-turn rotation, and glide reflection, but does not have horizontal reflection.

53. 12; see, for example, exercise 49(c).

G. Miscellaneous

55. **(a)** rotation; the two rigid motions that are proper are rotations and translations. Of these, only a rotation can have exactly one fixed point (the identity translation has infinitely many fixed points).

(b) identity motion; the two rigid motions that are proper are rotations and translations. Of these, only the identity motion has infinitely many fixed points.

(c) reflection; the two rigid motions that are improper are reflections and glide reflections. But glide reflections do not have fixed points while reflections can have an infinite number of them.

(d) glide reflection; the two rigid motions that are improper are reflections and glide reflections. Of these, only the glide reflection has no fixed points.

57. **(a)** D. The reflection of P about l_1 is the point E. The reflection of E about line l_2 is D. So, point D is the image of point P under the product of these reflections.

(b) D. The reflection of P about l_2 is the point A. The reflection of A about line l_1 is D. So, again, point D is the image of point P under the product of these reflections.

(c) B. The reflection of P about l_2 is the point A. The reflection of A about line l_3 is B.

(d) F. The reflection of P about l_3 is the point C. The reflection of C about line l_2 is F. [Note that (a), (b), (c) and (d) suggest that the order in which reflections occur *sometimes* makes a difference.]

(e) G. The reflection of P about l_1 is the point E. The reflection of E about line l_4 is G.

59. **(a)** When a proper rigid motion is combined with an improper rigid motion, the result is an improper rigid motion; that is, the left-right and clockwise-counterclockwise orientations on the final figure will be the reverse of the original figure.

(b) When an improper rigid motion is combined with an improper rigid motion, the result is a proper rigid motion; that is, the left-right and clockwise-counterclockwise orientations on the final figure will be the same as the original figure. This occurs because the orientations are reversed and then reversed again. The second reversal brings the orientation back to the original orientation.

(c) improper; this is an improper rigid motion combined with a proper rigid motion (see (a)).

(d) proper; this is an improper rigid motion combined with an improper rigid motion (see (b)).

61. The combination of two improper rigid motions is a proper rigid motion. Since *C* is a fixed point, the rigid motion must be a rotation with rotocenter *C*.

JOGGING

63. **(a)** The result of applying the reflection with axis l_1, followed by the reflection with axis l_2, is a clockwise rotation with center *C* and angle of rotation $\lambda + \lambda + \beta + \beta = 2(\lambda + \beta) = 2\alpha$. One example is shown in the figure.

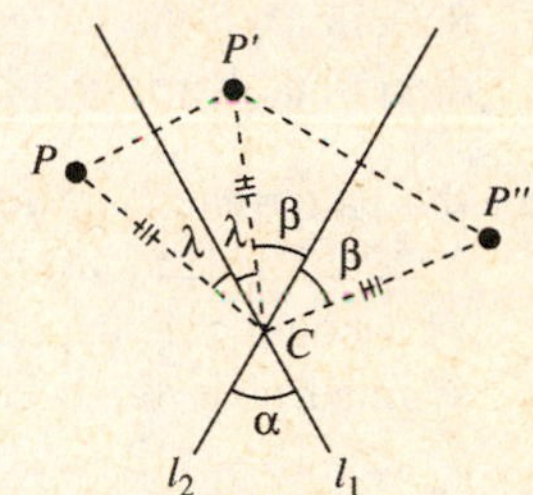

(b) The result of applying the reflection with axis l_2, followed by the reflection with axis l_1, is a *counter-clockwise* rotation with center *C* and angle of rotation 2α.

65. **(a)**

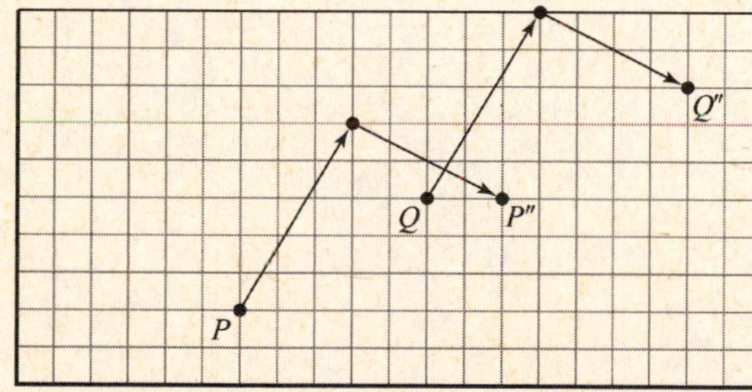

P" is the image of *P* under the product of $\mathcal{M}$ and $\mathcal{N}$.
Q" is the image of *Q* under the product of $\mathcal{M}$ and $\mathcal{N}$.

(b)

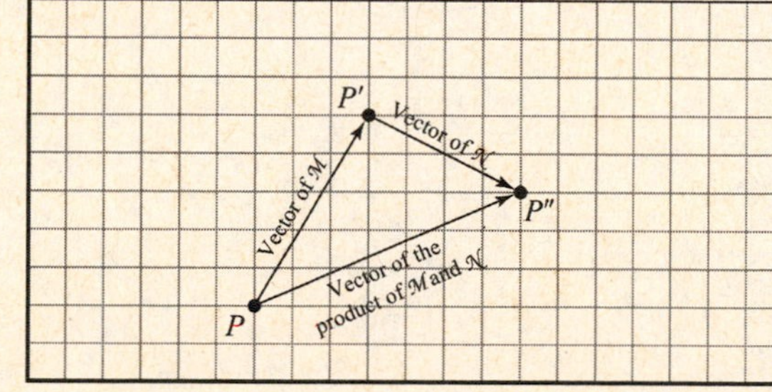

The vector of the translation representing the product of $\mathcal{M}$ and $\mathcal{N}$ is described by the arrow from *P* to *P*".

67. (a)

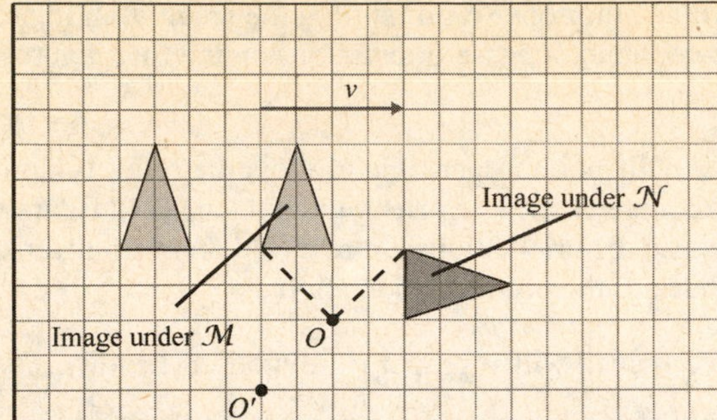

(b) Rotations and translations are proper rigid motions, and hence preserve clockwise-counterclockwise orientations. Thus, the product of $\mathcal{M}$ and $\mathcal{N}$ is a proper rigid motion. However, the given motion cannot be a translation since the vertices of the triangle are each moved different distances. The product is a $90°$ clockwise rotation about rotocenter O' (see figure in part (a)).

69. (a) By definition, a border has translation symmetries in exactly one direction (let's assume the horizontal direction). If the pattern had a reflection symmetry along an axis forming $45°$ with the horizontal direction, there would have to be a second direction of translation symmetry (vertical).

(b) If a pattern had a reflection symmetry along an axis forming an angle of $\alpha°$ with the horizontal direction, it would have to have translation symmetry in a direction that forms an angle of $2\alpha°$ with the horizontal. This could only happen for $\alpha = 90°$ or $\alpha = 180°$ (since the only allowable direction for translation symmetries is the horizontal).

71. Rotations and translations are proper rigid motions, and hence preserve clockwise-counterclockwise orientations. The given motion is an improper rigid motion (it reverses the clockwise-counterclockwise orientation). If the rigid motion was a reflection, then PP', RR', and QQ' would all be perpendicular to the axis of reflection and hence would all be parallel. It must be a glide reflection (the only rigid motion left).

Chapter 12

WALKING

A. The Koch Snowflake and Variations

1. (a)

	M	l	P
Start	3	1 cm	3 cm
Step 1	12	1/3 cm	4 cm
Step 2	48	1/9 cm	48/9 cm
Step 3	192	1/27 cm	192/27 cm
Step 4	768	1/81 cm	768/81 cm
Step 50	3×4^{50}	$1/3^{50}$ cm	53 km

At each step, the value of the number of sides M is multiplied by 4 and the value of each side length l is multiplied by 1/3. The perimeter P at each step is simply the product of M and l. The perimeter at Step 50 is thus $3\times4^{50}\times\frac{1}{3^{50}}=\left(\frac{4}{3}\right)^{50}\times3\approx5{,}297{,}343$ cm.

(b) M starts at a value of 3 and is multiplied by 4 at each step. So, at step N we have $M = 3\cdot4^N$;

l starts at a value of 1 cm and is multiplied by 1/3 at each step. So, at step N we have $l = \frac{1}{3^N}$ cm;

The perimeter P of the snowflake at step N is the product of M and l at step N, namely $P = 3\cdot\left(\frac{4}{3}\right)^N$ cm.

3. (a)

	R	S	T	Q
Start	0	0	0	81 in^2
Step 1	3	9 in^2	27 in^2	108 in^2
Step 2	12	1 in^2	12 in^2	120 in^2
Step 3	48	$\frac{1}{9}$ in^2	$\frac{16}{3}$ in^2	$\frac{376}{3}$ in^2
Step 4	192	$\frac{1}{81}$ in^2	$\frac{64}{27}$ in^2	$\frac{3448}{27}$ in^2
Step 5	768	$\frac{1}{729}$ in^2	$\frac{256}{243}$ in^2	$\frac{31228}{243}$ in^2

In the first step, 3 triangles are added. At each step thereafter, R (the number of triangles added) is multiplied by 4. The area S of each triangle added at a given step is 1/9 of the area added during the previous step. The total new area T added at a given step is then the product of R and S. The area of the "snowflake" obtained at a given step (which we call Q) is the sum of T and the area Q of the snowflake in the previous step.

For example, at step 1 a total of $R = 3$ triangles are added each having area $S = 9$ in^2 (S is 1/9 the area of the original triangle which was 81 in^2). So, the value of Q at step 1 is $81\text{ in}^2 + 3\times9\text{ in}^2 = 108\text{ in}^2$. At step 2, $R = 3\times4 = 12$ triangles are added each having area $S = 1$ in^2 (this can be seen as either 1/9 of the area added in step 1 or $1/9^2$ the area of the original triangle). The value of Q at step 2 is then $Q = 108\text{ in}^2 + 12\times 1\text{ in}^2 = 120\text{ in}^2$.

At step 3, $R = 3\times 4^2 = 48$ triangles are added each having area $S = 1/9\text{ in}^2$ (again, this can be seen as 1/9 of the previous value of S). The value of Q at step 3 is then $Q = 120\text{ in}^2 + 48\times\frac{1}{9}\text{ in}^2 = \frac{376}{3}\text{ in}^2$.

(b) $1.6\times 81\text{ in}^2 = 129.6\text{ in}^2$

5. (a)

	M	l	P
Start	4	9 cm	36 cm
Step 1	20	3 cm	60 cm
Step 2	100	1 cm	100 cm
Step 3	500	1/3 cm	500/3 cm
Step 30	4×5^{30}	$1/3^{28}$ cm	1628 km

At each step, the value of the number of sides M is multiplied by 5 and the value of each side length l is multiplied by 1/3. The perimeter P at each step is simply the product of M and l. The perimeter at Step 30 is thus $4\times 5^{30}\times\frac{1}{3^{28}} = 4\times 5^2\times\left(\frac{5}{3}\right)^{28} \approx 162{,}841{,}460$ cm.

(b) M starts at a value of 4 and is multiplied by 5 at each step. So, at step N we have $M = 4\cdot 5^N$;

l starts at a value of 9 cm and is multiplied by 1/3 at each step. At step N, $l = 9\cdot\frac{1}{3^N}\text{cm} = \frac{1}{3^{N-2}}$ cm;

The perimeter of the fractal at step N is the product of M and l at step N, namely

$$P = 4\cdot 5^N\cdot\left(\frac{1}{3}\right)^{N-2} = 4\cdot 5^2\cdot\left(\frac{5}{3}\right)^{N-2} = 100\cdot\left(\frac{5}{3}\right)^{N-2}\text{ cm.}$$

7.

	R	S	T	Q
Start	0	0	0	81
Step 1	4	9	36	117
Step 2	20	1	20	137
Step 3	100	$\frac{1}{9}$	$\frac{100}{9}$	$\frac{1333}{9}$
Step 4	500	$\frac{1}{81}$	$\frac{500}{81}$	$\frac{12497}{81}$

In the step 1, 4 squares are added. At each step thereafter, R (the number of squares added) is multiplied by 5. The area S of each square added at a given step is 1/9 of the area of each square added during the previous step. The total new area T added at a given step is then the product of R and S. The area of the shape obtained at a given step (which we call Q) is the sum of T and the area Q of the shape in the previous step.

At step 1, for example, a total of $R = 4$ squares are added each having area $S = 9$ (S is 1/9 the area of the

original square which was 81). So, the value of Q at step 1 is $81+4\times 9=117$.
At step 2, $R=4\times 5=20$ squares are added each having area $S = 1$ (this can be seen as 1/9 of the area added in step 1). The value of Q at step 2 is then $Q=117+20\times 1=137$.

At step 3, $R=4\times 5^2=100$ squares are added each having area $S = 1/9$ (again, this can be seen as 1/9 of the previous value of S). The value of Q at step 3 is then $Q=137+100\times\frac{1}{9}=\frac{1333}{9}$.

9. (a)

	M	l	P
Start	3	6 cm	18 cm
Step 1	12	2 cm	24 cm
Step 2	48	2/3 cm	32 cm
Step 3	192	2/9 cm	128/3 cm
Step 4	768	2/27 cm	512/9 cm
Step 30	3×4^{30}	$6/3^{30}$ cm	1 km

At each step, the value of the number of sides M is multiplied by 4 (each side becomes four) and the value of each side length l is multiplied by 1/3 (each side length is divided into three equal lengths). The perimeter P at each step is simply the product of M and l. The perimeter at Step 30 is thus

$$3\times 4^{30}\times\frac{6}{3^{30}}=2\times 4^2\times\left(\frac{4}{3}\right)^{28}\approx 100{,}793\text{ cm.}$$

(b) M starts at a value of 3 and is multiplied by 4 at each step. So, at step N we have $M = 3\cdot 4^N$;
l starts at a value of 6 cm and is multiplied by 1/3 at each step. At step N we have $l = 6\cdot\frac{1}{3^N}=\frac{2}{3^{N-1}}$ cm;
The perimeter of the fractal at step N is the product of M and l at step N, namely

$$P=3\cdot 4^N\cdot\left(\frac{1}{3}\right)^{N-1}\cdot 2=2\cdot 4^2\cdot\left(\frac{4}{3}\right)^{N-2}=32\cdot\left(\frac{4}{3}\right)^{N-2}\text{ cm.}$$

11. (a)

	R	S	T	Q
Start	0	0	0	81 in^2
Step 1	3	9 in^2	27 in^2	54 in^2
Step 2	12	1 in^2	12 in^2	42 in^2
Step 3	48	$\frac{1}{9}$ in^2	$\frac{48}{9}$ in^2	$\frac{110}{3}$ in^2
Step 4	192	$\frac{1}{81}$ in^2	$\frac{192}{81}$ in^2	$\frac{926}{27}$ in^2
Step 5	768	$\frac{1}{729}$ in^2	$\frac{768}{729}$ in^2	$\frac{8078}{243}$ in^2

In the first step, 3 triangles are subtracted. At each step thereafter, the number of triangles subtracted (which we call R) is multiplied by 4. The area S of each triangle subtracted at a given step is 1/9 of the area subtracted during the previous step. The total new area T subtracted at a given step is then the product of R and S. The area of the shape obtained at a given step (which we call Q) is the difference of the area of the shape in the previous step and T. [Note: This is similar to Exercise 3 except subtraction is used rather than addition.]

For example, at step 1 a total of $R = 3$ triangles are subtracted each having area $S = 9 \text{ in}^2$ (S is 1/9 the area of the original triangle which was 81 in^2). So, the value of Q at step 1 is $81 \text{ in}^2 - 3\times 9 \text{ in}^2 = 54 \text{ in}^2$. At step 2, $R = 3\times 4 = 12$ triangles are subtracted each having area $S = 1 \text{ in}^2$. The value of Q at step 2 is then $Q = 54 \text{ in}^2 - 12\times 1 \text{ in}^2 = 42 \text{ in}^2$.

At step 3, $R = 3\times 4^2 = 48$ triangles are subtracted each having area $S = 1/9 \text{ in}^2$ (again, this can be seen as 1/9 of the previous value of S). The value of Q at step 3 is then $Q = 42 \text{ in}^2 - 48\times \frac{1}{9} \text{ in}^2 = \frac{110}{3} \text{ in}^2$.

(b) $0.4\times 81 \text{ in}^2 = 32.4 \text{ in}^2$

13. (a)

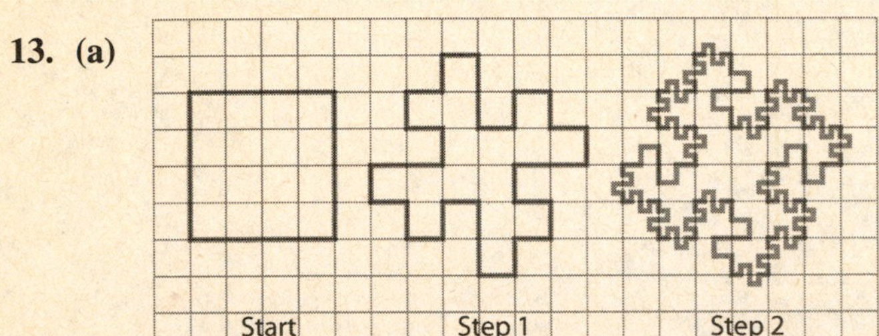

(b) The seed square has perimeter of $4\times 16 = 64$. By replacing each segment with the "sawtooth" version, the length is doubled. So, the perimeter of the entire figure obtained in step 1 is $2\times 64 = 128$.

(c) 256; the area at step 1 is the same as the area of the seed (what the rule "giveth," the rule "taketh" away).

(d) As in (b), replacing each segment with the "sawtooth" version doubles the length. So, since the perimeter in step 1 is 128, the perimeter of the figure obtained in step 2 is $2\times 128 = 256$.

(e) 256; again the area at step 2 is the same as the area at step 1 of construction.

15. At each step, the area added is the same as the area subtracted.

B. The Sierpinski Gasket and Variations

17.

	R	*S*	*T*	*Q*
Start	0	0	0	1
Step 1	1	1/4	1/4	3/4
Step 2	3	1/16	3/16	9/16
Step 3	9	1/64	9/64	27/64
Step 4	27	1/256	27/256	81/256
Step 5	81	1/1024	81/1024	243/1024

In the step 1, one triangle is subtracted. At each step thereafter, the number of triangles subtracted (R) is multiplied by 3. Starting with step 2, the area S of each triangle subtracted at a given step is 1/4 of the area of each triangle subtracted during the previous step (in step 1, the area of the triangle that is subtracted is 1/4). The total new area T subtracted at a given step is the product of R and S at that step. The area of the shape obtained at a given step (Q) is the difference between the area of the shape in the previous step and T.

For example, at step 3 shown in (d) in the figure below, $R = 1 \times 3^2 = 9$ triangles are subtracted each having area $S = 1/64$ (this can be seen as 1/4 of the previous value of S). So, T, the total area subtracted at step 3 is $9 \times 1/64 = 9/64$. The value of Q at step 3 is then $Q = \frac{9}{16} - \frac{9}{64} = \frac{27}{64}$.

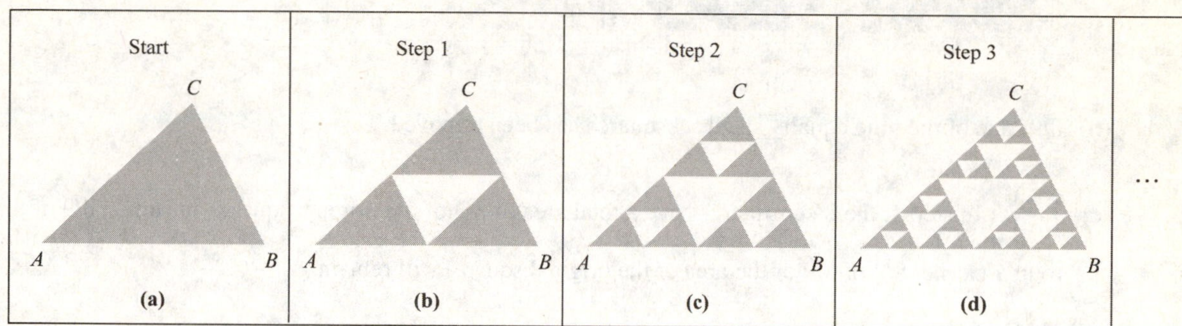

(a) (b) (c) (d)

19.

	U	V	W
Start	1	8 cm	8 cm
Step 1	3	4 cm	12 cm
Step 2	9	2 cm	18 cm
Step 3	27	1 cm	27 cm
Step 4	81	½ cm	81/2 cm
Step 5	243	¼ cm	243/4 cm

At each step, the value of the number of dark triangles U is tripled and the value of V (the perimeter of each dark triangle) is halved (since each side length is halved). The length of the boundary of the "gasket" W at each step is simply the product of U and V.

21. **(a)** From Exercise 18, we see that the value of R at step N is 3^{N-1}, the value of S at step N is $A/4^N$. So, the value of T at step N of the construction is $3^{N-1} \cdot \frac{A}{4^N} = \frac{A}{4} \cdot \left(\frac{3}{4}\right)^{N-1}$. It follows that Q, the area of the "gasket" obtained at step N of the construction is

$$A - \frac{A}{4} - \frac{A}{4} \cdot \left(\frac{3}{4}\right)^1 - \frac{A}{4} \cdot \left(\frac{3}{4}\right)^2 - \ldots - \frac{A}{4} \cdot \left(\frac{3}{4}\right)^{N-1} = A - \left(\frac{A}{4} + \frac{A}{4} \cdot \left(\frac{3}{4}\right)^1 + \frac{A}{4} \cdot \left(\frac{3}{4}\right)^2 + \ldots + \frac{A}{4} \cdot \left(\frac{3}{4}\right)^{N-1}\right)$$

$$= A - \frac{A}{4}\left(\frac{1 - \left(\frac{3}{4}\right)^N}{1 - \frac{3}{4}}\right) = A - A + A \cdot \left(\frac{3}{4}\right)^N = A \cdot \left(\frac{3}{4}\right)^N.$$

(b) The area of the Sierpinski gasket is smaller than the area of the gasket formed during any step of construction. That is, if the area of the original triangle is 1, then the area of the Sierpinski gasket is less than $\left(\frac{3}{4}\right)^N$ for every positive value of N. Since $0 < 3/4 < 1$, the value of $\left(\frac{3}{4}\right)^N$ can be made smaller than any positive quantity for a large enough choice of N. It follows that the area of the Sierpinski gasket can also be made smaller than any positive quantity.

23. (a)

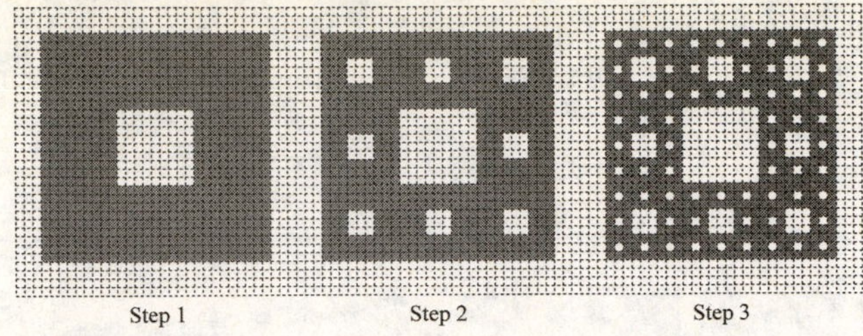

(b) 8/9; one of the nine equally sized subsquares has been removed.

(c) $\left(\frac{8}{9}\right)^2$; In step 1, the 8 subsquares have a total area of 8/9 of the original square. In step 2, 8/9 of this total area (i.e. 8/9 of 8/9 of the area of the original square) will remain.

(d) $\left(\frac{8}{9}\right)^3$

(e) $\left(\frac{8}{9}\right)^N$

25. (a) Note that the boundary will include the outside perimeter and the perimeters of all removed squares. We also note that each removed square will have sides of length 1/3 that of the square it was removed from. Since each side of the original square has length 1, the starting square has perimeter 4 and the length of the boundary of the "carpet" obtained in step 1 of the construction is $4+4\cdot\frac{1}{3}=4+\frac{4}{3}=\frac{12}{3}+\frac{4}{3}=\frac{16}{3}$. This represents the previous boundary length plus the boundary length of the new square hole with sides of length 1/3.

(b) In step 2 of the construction, the length of the boundary is $\frac{16}{3}+8\cdot\frac{4}{9}=\frac{80}{9}$. This is the length of the boundary in step 1 plus the perimeter of 8 new square holes, each having sides of length $(1/3)^2=1/9$.

(c) In step 3 of construction, the length of the boundary is $\frac{80}{9}+8^2\cdot\frac{4}{27}=\frac{496}{27}$. This is the length of the boundary at step 2 plus the perimeter of 8^2 new square holes, each having sides of length $(1/3)^3=1/27$.

27.

	R	*S*	*T*	*Q*
Start	0	0	0	1
Step 1	3	1/9	1/3	2/3
Step 2	18	1/81	2/9	4/9
Step 3	108	1/729	4/27	8/27
Step 4	648	1/6561	8/81	16/81
Step *N*	$3\cdot6^{N-1}$	$\frac{1}{9^N}$	$\frac{3\cdot6^{N-1}}{9^N}$	$\left(\frac{2}{3}\right)^N$

In step 1, three triangles are subtracted. In each step thereafter, the number of triangles subtracted (R) is multiplied by 6 (since six solid triangles remain for every three that are subtracted). Starting in step 2, the area S of each triangle subtracted at a given step is 1/9 of the area of each triangle subtracted during the previous step (in step 1, the area of each triangle that is subtracted is 1/9). The total new area T subtracted at a given step is the product of R and S at that step. The area of the shape obtained at a given step (Q) is the difference between the area of the shape in the previous step and T.

For example, in step 2 shown in the figure, $R = 3\times 6 = 18$ triangles are subtracted each having area $S = 1/81$ (this can be seen as 1/9 of the previous value of S). So, T, the total area subtracted in step 2 is $18\times 1/81 = 2/9$. The value of Q at step 2 is then $Q = \frac{2}{3} - \frac{2}{9} = \frac{4}{9}$.

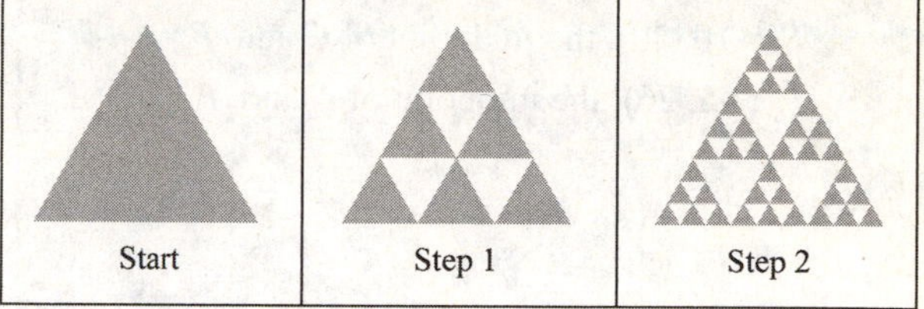

29.

	U	V	W
Start	1	9 cm	9 cm
Step 1	6	3 cm	18 cm
Step 2	36	1 cm	36 cm
Step 3	216	1/3 cm	72 cm
Step 4	1296	1/9 cm	144 cm
Step N	6^N	$1/3^{N-2}$ cm	$9\cdot 2^N$ cm

At each step, the value of the number of dark triangles U is multiplied by 6 and the value of V (the perimeter of each dark triangle) is 1/3 of that in the previous step (since each side length is also 1/3 of that in the previous step). The length of the boundary of the "ternary gasket" W at each step is simply the product of U and V.

31.

	Q
Start	A
Step 1	$3A/4$
Step 2	$9A/16$
Step 3	$27A/64$
Step 4	$81A/256$
Step N	$A\cdot \left(\frac{3}{4}\right)^N$

At each step, the area Q of the shape obtained at that step is 3/4 of the area of the shape formed in the previous step.

C. The Chaos Game and Variations

33. The coordinates of each point are:
P_1 : (32, 0);
P_2 : (16, 0), the midpoint of A and P_1;
P_3 : (8, 16), the midpoint of C and P_2;
P_4 : (20, 8), the midpoint of B and P_3;
P_5 :(10, 20), the midpoint of C and P_4;
P_6 : (5, 26), the midpoint of C and P_5.

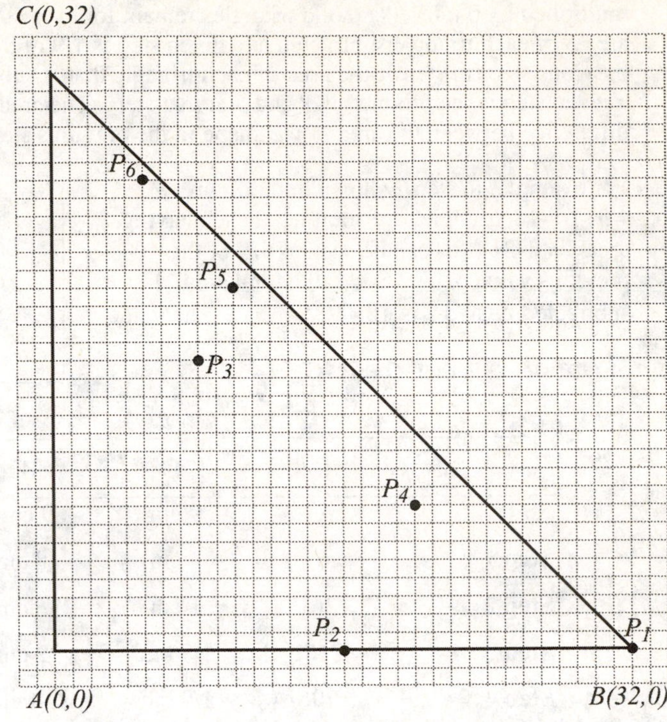

35.

Number rolled	Point	Coordinates
3	P_1	(32,0)
1	P_2	(16,0)
2	P_3	(8,0)
3	P_4	(20,0)
5	P_5	(10,16)
5	P_6	(5,24)

$P_3 = (8, 0)$ is the midpoint of A and P_2; $P_4 = (20,0)$ is the midpoint of B and P_3; $P_5 = (10, 16)$ is the midpoint of C and P_4; $P_6 = (5, 24)$, the midpoint of C and P_5.

37. General note: Find the new coordinate by picking the new x-value to be 2/3 of way from the x-coordinate of first point to x-coordinate of the second point and picking the new y-value to be 2/3 of the way from y-coordinate of the first point to y-coordinate of the second point.

(a) The coordinates of each point are:

P_1 : (0, 27);

P_2 : (18, 9), 2/3 of the way from P_1 to B;

P_3 : (6, 3), 2/3 of the way from P_2 to A;

P_4 : (20, 1), 2/3 of the way from P_3 to B.

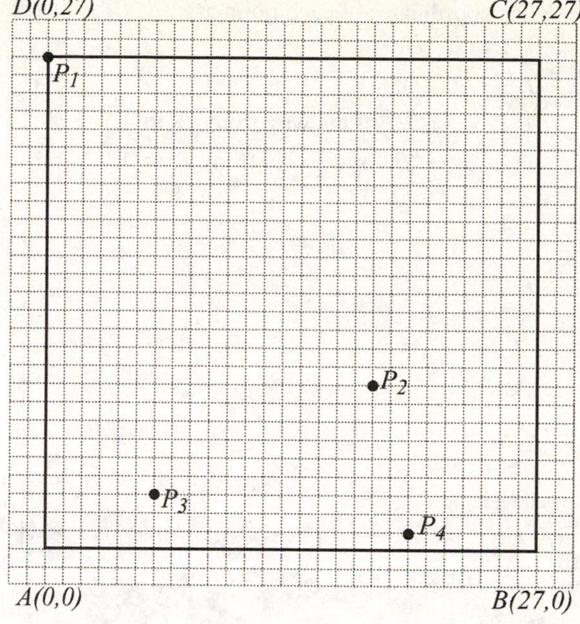

(b) The coordinates of each point are:

P_1 : (27, 27);

P_2 : (27, 9), 2/3 of the way from P_1 to B;

P_3 : (9, 3), 2/3 of the way from P_2 to A;

P_4 : (21, 1), 2/3 of the way from P_3 to B.

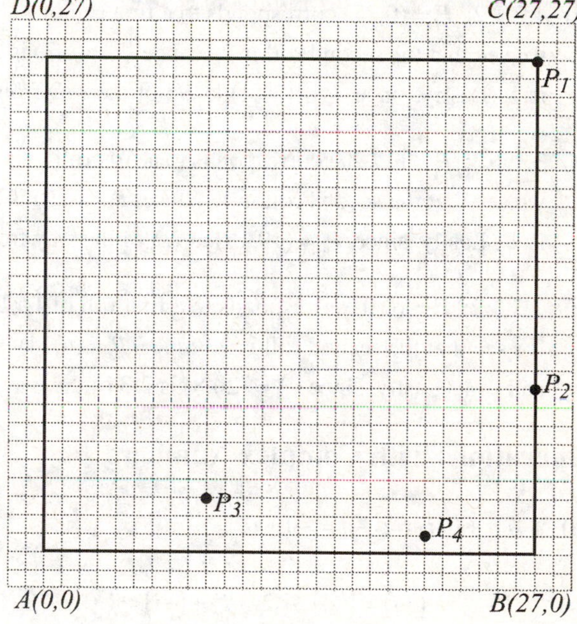

(c) The coordinates of each point are:
P_1 : (27, 27);
P_2 : (27, 27);
P_3 : (9, 9), 2/3 of the way from P_2 to *A*;
P_4 : (3, 3), 2/3 of the way from P_3 to *A*.

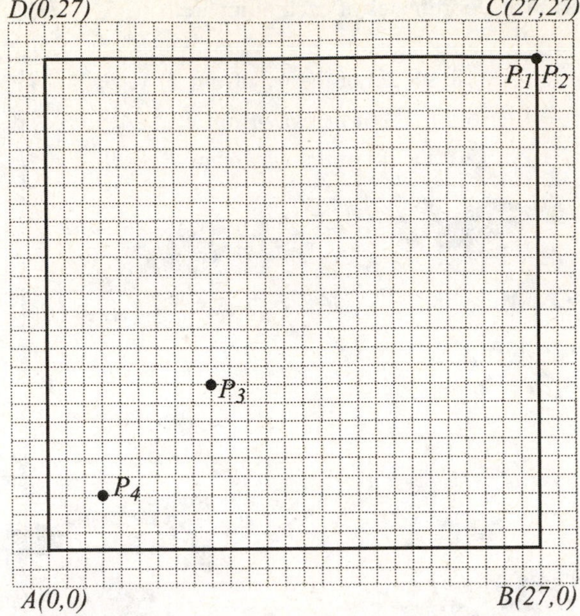

39. (a) 4,2,1,2; $P_1 = (0,27)$ corresponds to a first roll of 4. Then, since $P_2 = (18,9)$ is 2/3 of the way from P_1 to *B*, we see that the second roll is 2. Since $P_3 = (6,3)$ is 2/3 of the way from P_2 to *A*, we see that the third roll is 1. Lastly, since $P_4 = (20,1)$ is 2/3 of the way from P_3 to *B*, the fourth and final roll is a 2.

(b) 3,1,1,3; $P_1 = (27,27)$ corresponds to a first roll of 3. Then, since $P_2 = (9,9)$ is 2/3 of the way from P_1 to *A*, we see that the second roll is 1. Since $P_3 = (3,3)$ is 2/3 of the way from P_2 to *A*, we see that the third roll is 1. Lastly, since $P_4 = (19,19)$ is 2/3 of the way from P_3 to *C*, the fourth and final roll is a 3.

(c) 1,3,4,2; $P_1 = (0,0)$ corresponds to a first roll of 1. Then, since $P_2 = (18,18)$ is 2/3 of the way from P_1 to *C*, we see that the second roll is 3. Since $P_3 = (6,24)$ is 2/3 of the way from P_2 to *D*, we see that the third roll is 4. Lastly, since $P_4 = (20,8)$ is 2/3 of the way from P_3 to *B*, the fourth and final roll is a 2.

D. Operations with Complex Numbers

41. (a) $(-i)^2 + (-i) = i^2 - i = -1 - i$

(b) $(-1-i)^2 + (-i) = (-1)^2 + 2i + i^2 + (-i) = 1 + 2i - 1 - i = i$

(c) $i^2 + (-i) = -1 - i$

43. (a) $(-0.25+0.25i)^2 + (-0.25+0.25i) = 0.0625 - 0.125i - 0.0625 + (-0.25+0.25i) = -0.25 + 0.125i$

(b) $(-0.25-0.25i)^2 + (-0.25-0.25i) = 0.0625 + 0.125i - 0.0625 + (-0.25-0.25i) = -0.25 - 0.125i$

45. (a) Since the value of $i(1+i)=-1+i$, $i^2(1+i)=-1-i$, and $i^3(1+i)=1-i$ we plot $1+i$, $-1+i$, $-1-i$, and $1-i$. This is just like plotting (1,1), (-1,1), (-1,-1), and (1,-1) in the standard Cartesian plane.

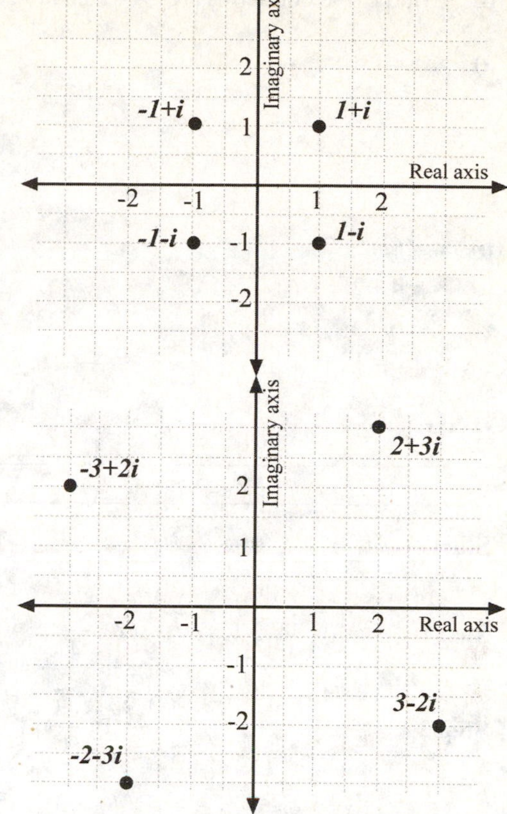

(b) Since the value of $i(3-2i)=2+3i$, $i^2(3-2i)=-3+2i$, and $i^3(3-2i)=-2-3i$ we plot $3-2i$, $2+3i$, $-3+2i$, and $-2-3i$.

(c) The effect of multiplying each point in (a) and (b) by i is a 90-degree counterclockwise rotation.

E. Mandelbrot Sequences

47. (a) $s_1=(-2)^2+(-2)=4-2=2$; $s_2=(2)^2+(-2)=4-2=2$; $s_3=(2)^2+(-2)=4-2=2$; $s_4=(2)^2+(-2)=4-2=2$.

(b) The sequence is attracted to 2 (in fact, each term equals 2), so $s_{100}=2$.

(c) Each number in the sequence is 2. In this case, the sequence could be considered periodic (with period 1) or attracted to 2 (in the sense that a sequence of 2's could be considered getting closer and closer to the attractor 2).

49. (a) $s_1=(-0.5)^2+(-0.5)=-0.25$; $s_2=(-0.25)^2+(-0.5)=-0.4375$; $s_3=(-0.4375)^2+(-0.5)\approx-0.3086$; $s_4=(-0.3086)^2+(-0.5)\approx-0.4048$; $s_5=(-0.4048)^2+(-0.5)\approx-0.3361$.

(b) $s_{N+1}=(-0.366)^2+(-0.5)\approx-0.366$

(c) From part (b), we see that $s_N=s_{N+1}$, so the sequence must be attracted to –0.3660 (rounded to 4 decimal places).

51. (a) For the first three values, see Exercise 41. $s_1=(-i)^2+(-i)=i^2-i=-1-i$;
$s_2=(-1-i)^2-i=1+2i+i^2-i=i$; $s_3=i^2-i=-1-i$; $s_4=(-1-i)^2-i=1+2i+i^2-i=i$;
$s_5=i^2-i=-1-i$.

(b) Periodic; the odd terms are $-1-i$ and the even terms are i.

53. **(a)** $s_{N+1} = (s_N)^2 + s$

$38 = (6)^2 + s$

$38 = 36 + s$

$s = 2$

(b) $s_1 = (2)^2 + 2$

$= 4 + 2$

$= 6;$

So, $N = 1$

JOGGING

55.

	R	S	T	Q
Start	0	0	0	$a^2\dfrac{\sqrt{3}}{4}$
Step 1	3	$\dfrac{1}{9}\left(a^2\dfrac{\sqrt{3}}{4}\right)$	$\dfrac{3}{9}\left(a^2\dfrac{\sqrt{3}}{4}\right)$	$\dfrac{4}{3}\left(a^2\dfrac{\sqrt{3}}{4}\right)$
Step 2	12	$\dfrac{1}{9^2}\left(a^2\dfrac{\sqrt{3}}{4}\right)$	$\dfrac{12}{9^2}\left(a^2\dfrac{\sqrt{3}}{4}\right)$	$\dfrac{40}{27}\left(a^2\dfrac{\sqrt{3}}{4}\right)$
Step 3	48	$\dfrac{1}{9^3}\left(a^2\dfrac{\sqrt{3}}{4}\right)$	$\dfrac{48}{9^3}\left(a^2\dfrac{\sqrt{3}}{4}\right)$	$\dfrac{376}{243}\left(a^2\dfrac{\sqrt{3}}{4}\right)$
Step 4	192	$\dfrac{1}{9^4}\left(a^2\dfrac{\sqrt{3}}{4}\right)$	$\dfrac{192}{9^4}\left(a^2\dfrac{\sqrt{3}}{4}\right)$	$\dfrac{3448}{2187}\left(a^2\dfrac{\sqrt{3}}{4}\right)$
Step *N*	$3\cdot 4^{N-1}$	$\dfrac{1}{9^N}\left(a^2\dfrac{\sqrt{3}}{4}\right)$	$\dfrac{3\cdot 4^{N-1}}{9^N}\left(a^2\dfrac{\sqrt{3}}{4}\right)$	$\left(a^2\dfrac{\sqrt{3}}{4}\right)+\dfrac{3}{5}\cdot\left(a^2\dfrac{\sqrt{3}}{4}\right)\left[1-\left(\dfrac{4}{9}\right)^N\right]$

In step 1, three triangles are added. At each step thereafter, the number of triangles added (R) is multiplied by 4. The area S of each triangle added at a given step is 1/9 of the area added during the previous step. The total new area T added at a given step is the product of R and S. The area of the "snowflake" obtained at a given step (which we denote by Q) is the sum of T and the area Q of the snowflake in the previous step.

In step N, the value of Q can be found by either using the formula found in the middle of page 442 or by applying the geometric sum formula as shown below.

$$\left(a^2\frac{\sqrt{3}}{4}\right)+\frac{3}{9}\left(a^2\frac{\sqrt{3}}{4}\right)+\frac{3\cdot 4}{9^2}\left(a^2\frac{\sqrt{3}}{4}\right)+\frac{3\cdot 4^2}{9^3}\left(a^2\frac{\sqrt{3}}{4}\right)+\frac{3\cdot 4^3}{9^4}\left(a^2\frac{\sqrt{3}}{4}\right)+\ldots+\frac{3\cdot 4^{N-1}}{9^N}\left(a^2\frac{\sqrt{3}}{4}\right)=$$

$$\left(a^2\frac{\sqrt{3}}{4}\right)+\left(a^2\frac{\sqrt{3}}{4}\right)\left[\frac{3}{9}+\frac{3\cdot 4}{9^2}+\frac{3\cdot 4^2}{9^3}+\frac{3\cdot 4^3}{9^4}+\ldots+\frac{3\cdot 4^{N-1}}{9^N}\right]=$$

$$\left(a^2\frac{\sqrt{3}}{4}\right)+\frac{3}{9}\left(a^2\frac{\sqrt{3}}{4}\right)\left[1+\frac{4}{9}+\left(\frac{4}{9}\right)^2+\left(\frac{4}{9}\right)^3+\ldots+\left(\frac{4}{9}\right)^{N-1}\right]=$$

$$\left(a^2\frac{\sqrt{3}}{4}\right)+\frac{3}{9}\left(a^2\frac{\sqrt{3}}{4}\right)\left[\frac{1-\left(\frac{4}{9}\right)^N}{1-\frac{4}{9}}\right]=\left(a^2\frac{\sqrt{3}}{4}\right)+\frac{3}{5}\left(a^2\frac{\sqrt{3}}{4}\right)\left[1-\left(\frac{4}{9}\right)^N\right]$$

57. (a)

	C	U	V
Start	0	0	1
Step 1	7	1/27	20/27
Step 2	20×7	$\left(\frac{1}{27}\right)^2$	$\left(\frac{20}{27}\right)^2$
Step 3	$20^2\times 7$	$\left(\frac{1}{27}\right)^3$	$\left(\frac{20}{27}\right)^3$
Step 4	$20^3\times 7$	$\left(\frac{1}{27}\right)^4$	$\left(\frac{20}{27}\right)^4$

At the first step, a cube is removed from each of the six faces and from the center for a total of $C = 7$ cubes removed. At the second step, each of the 20 remaining cubes has 7 cubes removed (i.e, $C = 20\times 7$ cubes are removed). In step 3, there are 20^2 remaining cubes each of which has 7 cubes removed (i.e. $C = 20^2\times 7$ cubes are removed). Etc.

Since step 1 can be thought of as $3\times 3\times 3 = 27$ equally sized cubes, the volume U of the middle cube removed is 1/27. For step 2 and onward, the volume U of each cube removed is 1/27 the volume of each cube removed in the previous step. The volume V of the sponge at any particular step is simply the difference of the previous value of V (the volume at the previous step) and the product of C and U (the volume removed during the current step).

(b) Since $20/27 < 1$, the Menger Sponge has infinitesimally small volume. (If a number between 0 and 1 is multiplied by itself over and over again, the resulting product will be close to 0.)

59. (a) Reflection with axis a vertical line passing through the center of the snowflake; reflections with axes lines making 30°, 60°, 90°, 120°, 150° angles with the vertical axis of the snowflake and passing through the center of the snowflake.

(b) Rotations of 60°, 120°, 180°, 240°, 300°, 360° with rotocenter the center of the snowflake.

(c) D_6

61. (a) Using the formula for the sum of the terms in a geometric sequence in which $r = 4/9$ (found in chapter 10), we have $1+\left(\frac{4}{9}\right)+\left(\frac{4}{9}\right)^2+\ldots+\left(\frac{4}{9}\right)^{N-1}=\dfrac{1-\left(\frac{4}{9}\right)^N}{1-\left(\frac{4}{9}\right)}=\frac{9}{5}\cdot\left[1-\left(\frac{4}{9}\right)^{N-1}\right]$.

(b) Using the result in (a) we have $\left(\frac{A}{3}\right)+\left(\frac{A}{3}\right)\left(\frac{4}{9}\right)+\left(\frac{A}{3}\right)\left(\frac{4}{9}\right)^2+\ldots+\left(\frac{A}{3}\right)\left(\frac{4}{9}\right)^{N-1}=$
$\left(\frac{A}{3}\right)\left[1+\left(\frac{4}{9}\right)+\left(\frac{4}{9}\right)^2+\ldots+\left(\frac{4}{9}\right)^{N-1}\right]=\left(\frac{A}{3}\right)\left(\frac{9}{5}\right)\left[1-\left(\frac{4}{9}\right)^N\right]=\left(\frac{3}{5}\right)A\left[1-\left(\frac{4}{9}\right)^N\right].$

63. If it is attracted, there will be a solution to $s_{N+1}=(s_N)^2+0.25$ with $s_{N+1}=s_N$. Substituting, we have $s_N=s_N{}^2+0.25$ or $s_N{}^2-s_N+0.25=0$. Solving via the quadratic equation, $s_N=\dfrac{1\pm\sqrt{1-4\left(\frac{1}{4}\right)}}{2}=\dfrac{1\pm 0}{2}=\dfrac{1}{2}$.
Examining several terms of the sequence (see below) indicates that the sequence is indeed attracted to 1/2.
$s_1=0.25^2+0.25=0.3125$; $s_2=0.3125^2+0.25\approx 0.3477$; $s_3=0.3477^2+0.25\approx 0.3709$; … $s_{30}\approx 0.4725$.

65. Consider the first term of the sequence: $s_1=\left(\sqrt{2}\right)^2+\sqrt{2}=2+\sqrt{2}$. Since $s_1>1$ and $\sqrt{2}>0$, we are guaranteed that $s_{N+1}>s_N$ and so the sequence is escaping. Investigating several terms will also support this reasoned conclusion.

Mini-Excursion 3

A. Linear Growth and Arithmetic Sequences

1. (a) $P_{30} = 75 + 5 \times 30 = 225$

(b) $1000 \le 75 + 5N$

$925 \le 5N$

$N \ge 185$

(c) The population will reach (and surpass) 1002 in the generation after it reaches 1000. Thus, it will take 186 generations for the population to reach 1002.

3. (a) $38 = 8 + 10 \times d$

$30 = 10d$

$d = 3$

(b) $P_{50} = 8 + 50 \times 3 = 158$

(c) $P_N = 8 + 3N$

5. (a) $2 + 5 \times 99 = 497$

(b) $A_0 = 2,\ d = 5$

The 100th term is $A_{99} = 2 + 5 \times 99 = 497$.

This is the sum of 100 terms in an arithmetic sequence:

$$2 + 7 + 12 + \ldots + 497 = \frac{(2 + 497) \times 100}{2} = 24{,}950$$

7. (a) $d = 3$

If 309 is the Nth term of the sequence,

$309 = 12 + 3(N - 1)$

$309 = 12 + 3N - 3$

$300 = 3N$

$N = 100$

So, 309 is the 100th term of the sequence.

(b) $$12 + 15 + 18 + \ldots + 309 = \frac{(12 + 309) \times 100}{2} = 16{,}050$$

9. (a) $P_{38} = 137 + 2 \times 38 = 213$

(b) $P_N = 137 + 2N$

(c) $137 \times \$1 \times 52 = \$7124$

(d) When just counting the newly installed lights, $P_0 = 0$ and $P_{51} = 2 \times 51 = 102$. (The lights installed in the 52nd week aren't in operation during the 52-week period.)

$$0 + 2 + 4 + \ldots + 102 = \frac{(0 + 102) \times 52}{2} = 2652$$

The cost is \$2,652.

B. Exponential Growth and Geometric Sequences

11. (a) $P_1 = 11 \times 1.25 = 13.75$

(b) $P_9 = 11 \times (1.25)^9 \approx 81.956$

(c) $P_N = 11 \times (1.25)^N$

13. (a) $P_N = 1.50 P_{N-1}$ where $P_0 = 200$

We multiply by 1.50 each year to account for an *increase* by 50%.

(b) $P_N = 200 \times 1.5^N$

(c) $P_{10} = 200 \times 1.5^{10} \approx 11{,}533$

A good estimate would be to say that about 11,500 crimes will be committed in 2019.

C. Logistic Growth Model

15. (a) $p_1 = 0.8 \times (1 - 0.3) \times 0.3 = 0.1680$

(b) $p_2 = 0.8 \times (1 - 0.168) \times 0.168 \approx 0.1118$

(c) $p_3 = 0.8 \times (1 - 0.11182) \times 0.11182 \approx 0.07945$

Thus, approximately 7.945% of the habitat's carrying capacity is taken up by the third generation.

17. (a) Using the formula $p_{N+1} = r(1 - p_N)p_N$ and a calculator with a memory register or a spreadsheet, we get

$p_1 = 0.1680,\ p_2 \approx 0.1118,\ p_3 \approx 0.0795,$

$p_4 \approx 0.0585,\ p_5 \approx 0.0441,\ p_6 \approx 0.0337,$

$p_7 \approx 0.0261,\ p_8 \approx 0.0203,\ p_9 \approx 0.0159,$

$p_{10} \approx 0.0125.$

(b) Since $p_N \to 0$ this logistic growth model predicts extinction for this population.

19. **(a)** Using the formula $p_{N+1} = r(1-p_N)p_N$ and a calculator with a memory register or a spreadsheet, we get $p_1 = 0.4320$, $p_2 \approx 0.4417$, $p_3 \approx 0.4439$, $p_4 \approx 0.4443$, $p_5 \approx 0.4444$, $p_6 \approx 0.4444$, $p_7 \approx 0.4444$, $p_8 \approx 0.4444$, $p_9 \approx 0.4444$, $p_{10} \approx 0.4444$.

(b) The population becomes stable at $\frac{4}{9} \approx 44.44\%$ of the habitat's carrying capacity.

21. **(a)** $p_1 = 0.3570$, $p_2 \approx 0.6427$, $p_3 \approx 0.6429$, $p_4 \approx 0.6428$, $p_5 \approx 0.6429$, $p_6 \approx 0.6428$, $p_7 \approx 0.6429$, $p_8 \approx 0.6428$, $p_9 \approx 0.6429$, $p_{10} \approx 0.6428$

(b) The population becomes stable at $\frac{9}{14} \approx 64.29\%$ of the habitat's carrying capacity.

23. **(a)** $p_1 = 0.5200$, $p_2 = 0.8112$, $p_3 \approx 0.4978$, $p_4 \approx 0.8125$, $p_5 \approx 0.4952$, $p_6 \approx 0.8124$, $p_7 \approx 0.4953$, $p_8 \approx 0.8124$, $p_9 \approx 0.4953$, $p_{10} \approx 0.8124$

(b) The population settles into a two-period cycle alternating between a high-population period at approximately 81.24% and a low-population period at approximately 49.53% of the habitat's carrying capacity.

D. Miscellaneous

25. **(a)** Exponential ($r = 2$)

(b) Linear ($d = 2$)

(c) Logistic

(d) Exponential ($r = \frac{1}{3}$)

(e) Logistic

(f) Linear ($d = -0.15$)

(g) Linear ($d = 0$), Exponential ($r = 1$), and/or Logistic (they all apply!)

27. The first N terms of the arithmetic sequence are $c, c+d, c+2d, \ldots c+(N-1)d$. Their sum is $\frac{(c+[c+(N-1)d])\times N}{2} = \frac{N}{2}[2c+(N-1)d]$.

29. No. This would require $p_0 = p_1 = 0.8(1-p_0)p_0$ and $p_0 = 0$ or $1 = 0.8(1-p_0)$. So, $p_0 = 0$ or $p_0 = -0.25$, neither of which are possible.

Chapter 13

WALKING

A. Surveys and Polls

1. (a) The population in this study is the collection of objects that are under study. For this survey, it is the gumballs in the jar.

   (b) A sample is a subgroup of a population from which data is collected. For this survey, it is the 25 gumballs drawn out of the jar.

   (c) 32%; The proportion of red gumballs in the sample is $\frac{8}{25} = 0.32$.

   (d) The sampling method used for this survey was simple random sampling. Each set of *n* gumballs had the same probability of being selected as any other set of *n* gumballs.

3. (a) 25%; The parameter for the proportion of red gumballs in the jar is $\frac{50}{200} = 0.25$.

   (b) The sampling error, the difference between the parameter and the statistic, is 32% - 25% = 7%.

   (c) Sampling variability. Simple random sampling was used. This survey method does not suffer from selection bias.

5. (a) The sampling proportion for this survey is $\frac{680}{8325} \approx 0.082$ (approximately 8.2%).

   (b) 306/680 = 45%; of those sampled, this represents the percentage that indicated they planned to vote for Smith.

7. Of the people surveyed, 45% indicated they would vote for Smith $\left(\frac{306}{680} = 0.45\right)$. The actual percentage was 42%. Since 45% – 42% = 3%, the sampling error for Smith was 3%. Similarly, the sampling error for Jones was 43% - 40% = 3% and the sampling error for Brown was 15% - 15% = 0%.

9. (a) The sample for this survey is the 350 students attending the Eureka High football game before the election.

   (b) $\frac{350}{1250} = 28\%$

11. (a) The population consists of all 1250 students at Eureka High School while the sampling frame consists only of those students that attended the football game a week prior to the election.

    (b) The sampling error is mainly a result of sampling bias. The sampling frame (and hence any sample taken from it) is not representative of the population. Students that choose to attend a football game are not representative of all Eureka High students.

13. (a) The sampling frame is all married people who read Dear Abby's column.

    (b) Abby's target population appears to be all married people. However, she is sampling from a (non-representative) subset of the population – a sampling frame that consists of those married couples that read her column. The sampling frame is quite different than the target population.

    (c) The sample chose itself. That is, the sample was chosen via self-selection (which is a type of convenience sample).

    (d) 85% is a statistic, since it is based on data taken from a sample.

15. (a) $\frac{44,807}{60,550} = 74.0\%$

    (b) $\frac{127,318 + 44,807}{210,336} \approx 81.8\%$

    (c) These estimates are probably not very accurate. The sample was far from being representative of the entire target population. One reason is that the sampling frame is so different from the target population. Another reason is that even if the sampling frame was similar to the target population, the survey is subject to nonresponse bias.

17. (a) The target population of this survey is the citizens of Cleansburg.

(b) The sampling frame is limited to that part of the target population that passes by a city street corner between 4:00 p.m. and 6:00 p.m.. It excludes citizens of Cleansburg having other responsibilities during that time of day.

19. (a) The choice of street corner could make a great deal of difference in the responses collected.

(b) *D* (We are making the assumption that people who live or work downtown are much more likely to answer yes than people in other parts of town.)

(c) Yes, the survey was subject to selection bias for two main reasons. (i) People out on the street between 4 p.m. and 6 p.m. are not representative of the population at large. For example, office and white-collar workers are much more likely to be in the sample than homemakers and school teachers. (ii) The five street corners were chosen by the interviewers and the passersby are unlikely to represent a cross section of the city.

(d) No, no attempt was made to use quotas to get a representative cross section of the population.

21. (a) Assuming that the registrar has a complete list of the 15,000 undergraduates at Tasmania State University, the target population and the sampling frame both consist of all undergraduates at TSU.

(b) $N = 15{,}000$

23. (a) In simple random sampling, any two members of the population have as much chance of both being in the sample as any other two. But in this sample, two people with the same last name—say Len Euler and Linda Euler—have no chance of being in the sample together.

(b) Sampling variability. The students sampled appear to be a reasonably representative cross section of all TSU undergraduates that might or might not be familiar with the new financial aid program.

25. (a) Stratified sampling. The trees are broken into three different strata (by variety) and then a random sample is taken from each stratum.

(b) Quota sampling. The grower is using a systematic method to force the sample to fit a particular profile. However, because the grower is human, sampling bias could then be introduced. When selecting 300 trees of variety A, the grower does not select them at random. Selecting 300 trees in one particular part of the orchard could bias the yield.

27. (a) Convenience sampling.
George is selecting those units in the population that are easily accessible..

(b) Stratified sampling.
The student newspaper is dividing the population into strata and then selecting a proportionately sized random sample from each stratum (in fact, 5% from each stratum).

(c) Simple random sampling.
Every subset of three players has the same chance of being selected as any other subset of three players.

(d) Quota sampling.
The coach is attempting to force the sample to fit a particular profile.

29. (a) Label the crates with numbers 1 to 250. Select 20 of these numbers at random (put the 250 numbers in a hat and draw out 20). Sample those 20 crates. The important point is that *any* set of 20 crates have an equal chance of being in the sample.

(b) Select the top 20 crates in the shipment (those easiest to access).

(c) Randomly sample 6 crates from supplier A, 6 crates from supplier B, and 8 crates from supplier C.

(d) Select any 6 crates from supplier A, any 6 crates from supplier B, and any 8 crates from supplier C.

B. The Capture-Recapture Method

31. $N = \frac{n_2}{k} \cdot n_1 = \frac{120}{30} \cdot 500 = 2000$

33. $N = \frac{n_2}{k} \cdot n_1 = \frac{28}{4} \cdot 36 = 252$ quarters.
(Remember: To estimate the number of quarters, we disregard the nickels and dimes—they are irrelevant.)

35. $N = \frac{n_2}{k} \cdot n_1 = \frac{43}{8} \cdot 69 = 370.875$; Rounding to the nearest integer gives us 371 dimes. Since 252 quarters, 261 nickels, and 371 dimes are estimated to be in the jar (see exercises 33 and 34), the total amount of money in the jar is estimated to be
$252 \times (\$0.25) + 261 \times (\$0.05) + 371 \times (\$0.10) =$
$\$63.00 + \$13.05 + \$37.10 = \$113.15$

37. Using $\frac{k}{n_2} = \frac{n_1}{N}$ with $k = 7, n_1 = 1700$, and $N =$ 160,000, we solve $\frac{7}{n_2} = \frac{1700}{160,000}$ for n_2 to find $n_2 \approx 660$.

39. Capture-recapture would underestimate the true population. If the fraction of those tagged in the recapture appears higher than it is in reality, then the fraction of those tagged in the initial capture will also be computed as higher than the truth. This makes the total population appear smaller than it really is.

For an example to see how this works, suppose a population had $N = 1000$ individuals. Imagine an initial tagging of $n_1 = 100$ individuals (10% of the population are tagged). Then, in the recapture of $n_2 = 20$, we should expect $k = 2$ individuals are tagged. If, however, some of the individuals were "trap-happy," then the number of individuals that are captured twice (the value of k) would be higher than expected (more than 2). Say $k = 4$ individuals (double the original number) were captured twice. Then, the value of N would be estimated to be
$N = \frac{n_2}{k} \cdot n_1 = \frac{20}{4} \cdot 100 = 500$. This is clearly an underestimate (half the original number).

C. Clinical Studies

41. **(a)** The target population for this study is anyone who could have a cold and would consider buying vitamin X (i.e., pretty much all adults).

(b) The sampling frame is only a small portion of the target population. It only consists of college students in the San Diego area that are suffering from colds. The sample is 500 of these San Diego area students all of whom are suffering from a cold at the start of the study.

(c) Yes. This sample would likely under represent older adults and those living in colder climates.

43. Four different problems with this study that indicate poor design include
(i) using college students (College students are not a representative cross section of the population in terms of age and therefore in terms of how they would respond to the treatment.),
(ii) using subjects only from the San Diego area,
(iii) offering money as an incentive to participate, and
(iv) allowing self-reporting (the subjects themselves determine when their colds are over).

45. The target population for this study is all potential knee surgery patients.

47. **(a)** Yes, this was a controlled placebo experiment. There was one control group receiving the sham-surgery (placebo) and two treatment groups.

(b) The first treatment group consisted of those patients receiving arthroscopic debridement. The second treatment group consisted of those patients receiving arthroscopic lavage.

(c) Yes, this study could be considered a randomized controlled experiment since the 180 patients in the study were assigned to a treatment group at random.

(d) This was a blind experiment. The doctors certainly knew which surgery they were performing on each patient.

49. The target population consists of all people having a history of colorectal adenomas. The point of the study is to determine the effect that Vioxx had on this population. That is, whether recurrence of colorectal polyps could be prevented in this population.

51. **(a)** The treatment group consisted of the 1287 patients that were given 25 daily milligrams of Vioxx. The control group consisted of the 1299 patients that were given a placebo.

(b) This is an experiment since members of the population received a treatment. It is a controlled placebo experiment since there was a control group that did not receive the treatment, but instead received a placebo. It is a randomized controlled experiment since the 2586 participants were randomly divided into the treatment and control groups. The study was double blind since neither the participants nor the doctors involved in the clinical trial knew who was in each group.

53. **(a)** women; particularly young women

(b) The sampling frame consists of those women between 16 and 23 years of age that are not at high risk for HPV infection (that is, those women having no prior abnormal pap smears and at most five previous male sexual partners). Pregnant women are also excluded from the sampling frame due to the risks involved. Though not perfect, it appears to be a representative subset of the target population.

55. **(a)** The treatment group consists of the half of the 2392 women in the sample that received the HPV vaccine.

(b) This is a controlled placebo experiment since there was a control group that did not receive the treatment, but instead received a placebo injection. It is a randomized controlled experiment since the 2392 participants were randomly divided into the treatment and control groups. The study was likely double blind – it is probable that neither the participants nor the medical personnel giving the injections knew who was in each group.

57. **(a)** The stated purpose of the study was to determine the effectiveness of a new serum for treating diphtheria. So, the target population consisted of all people suffering from diphtheria.

(b) The sampling frame consists of those people having diphtheria symptoms serious enough to be admitted to one particular Copenhagen hospital (between May 1896 and May 1897). It appears to be a reasonably representative subset of the target population. This is primarily because there is likely little difference between those suffering from diphtheria in this Copenhagen hospital and those suffering from diphtheria elsewhere.

59. **(a)** The treatment group consisted of those patients admitted on the "even" days that received both the new serum and the standard treatment. The control group consisted of those patients admitted on the "odd" days that received only the standard treatment for diphtheria at the time.

(b) When received together with standard treatment, this new serum appears to be somewhat effective at treating diphtheria. Significantly more patients that did not receive the treatment in this study died (presumably many of them from diphtheria or related issues).

D. Miscellaneous

61. **(a)** Spurlock's study was a clinical trial since a treatment was imposed (eating three meals at McDonald's every day for 30 days) on a sample of the population.

(b) The target population is the set of "average Americans."

(c) The sample consisted of one person (that being Mr. Spurlock himself).

(d) Three problems with this study that indicate poor design include the following (there are countless others).
(i) The use of a sample that is not representative of the population.
(ii) A small sample size (1 person).
(iii) The lack of a control group in which a sample of "average Americans" curtailed their physical activity and ate the same number of calories as the treatment group.

63. **(a)** This study was a clinical trial since a treatment was imposed (the heavy reliance on supplemental materials, online practice exercises and interactive tutorials) on a sample of the population.

(b) Some possible confounding variables in this study include the following.
(i) The instructors used in the treatment group may be more excited about this new curricular approach or they may be better teachers.
(ii) If students in this particular intermediate algebra class were self-selected, they may not be representative of the target population.
(iii) Students in the treatment group may have benefited simply by being forced to put more time studying into the course. To eliminate this possible confounding variable, the control groups should be asked to spend the same amount of time studying out of class.

65. (a) parameter
This statement refers to the entire population of students taking the SAT math test.

(b) statistic
A sample of the population of new automobiles are crash tested.

(c) statistic
A sample of the population of Mr. Johnson's blood tested positive.

(d) statistic
The poll did not sample all Americans; the statement refers to a sample of the population.

JOGGING

67. (a) The populations are (i) the entire sky; (ii) all the coffee in the cup; (iii) the entire Math 101 class.

(b) In none of the three examples is the sample random.

(c) (i) In some situations one can have a good idea as to whether it will rain or not by seeing only a small section of the sky, but in many other situations rain clouds can be patchy and one might draw the wrong conclusions by just peeking out the window. (ii) If the coffee is burning hot on top, it is likely to be pretty hot throughout, so Betty's conclusion is likely to be valid. (iii) Since Carla used convenience sampling and those students sitting next to her are not likely to be a representative sample, her conclusion is likely to be invalid.

69. (a) The results of this survey might be invalid because the question was worded in a way that made it almost impossible to answer yes.

(b) "Will you support some form of tax increase if it can be proven to you that such a tax increase is justified?" is better, but still not neutral. "Do you support or oppose some form of tax increase?" is bland but probably as neutral as one can get.

(c) Many such examples exist. A very real example is a question such as "Would you describe yourself as pro-life or not?" or "Would you describe yourself as pro-choice or not?"

71. (a) Under method 1, people whose phone numbers are unlisted are automatically ruled out from the sample. At the same time, method 1 is cheaper and easier to implement than method 2. Method 2 will typically produce more reliable data since the sample selected would better represent the population.

(b) For this particular situation, method 2 is likely to produce much more reliable data than method 1. The two main reasons are (i) people with unlisted phone numbers are very likely to be the same kind of people that would seriously consider buying a burglar alarm, and (ii) the listing bias is more likely to be significant in a place like New York City. (People with unlisted phone numbers make up a much higher percentage of the population in a large city such as New York than in a small town or rural area. Interestingly enough, the largest percentage of unlisted phone numbers for any American city is in Las Vegas, Nevada.)

73. (a) Fridays, Saturdays, and Sundays make up 3/7 or about 43% of the week. It follows that proportionally, there are fewer fatalities due to automobile accidents on Friday, Saturday, and Sunday (42%) that there are on Monday through Thursday.

(b) On Saturday and Sunday there are fewer people commuting to work. The number of cars on the road and miles driven is significantly less on weekends. The number of accidents due to fatalities should be proportionally much less. The fact that it is 42% indicates that there are other factors involved and increased drinking is one possible explanation.

75. **(a)** People are very unlikely to tell IRS representatives that they cheated on their taxes even when they are promised confidentiality.

(b) The critical issue in this survey is to get the respondents to given an honest answer to the question. In this regard two points are critical: (i) the survey is much more likely to get honest responses when sponsored by a "neutral" organization (a newspaper poll, for example); (ii) a mailed questionnaire is the safest way to guarantee anonymity for the respondent and in this case it is much better than a telephone or (heaven forbid) personal interview. This is a situation where a tradeoff must be made — some nonresponse bias is still preferable to dishonest answers. The nonresponse bias can be reduced by the use of an appropriate inducement or reward to those who can show proof of having mailed back the questionnaire (for example by showing a post office receipt, etc.).

Chapter 14

WALKING

A. Tables, Bar Graphs, Pie Charts, and Histograms

1.

Score	10	50	60	70	80	100
Frequency	1	3	7	6	5	2

3. (a)

Grade	A	B	C	D	F
Frequency	7	6	7	3	1

(b)

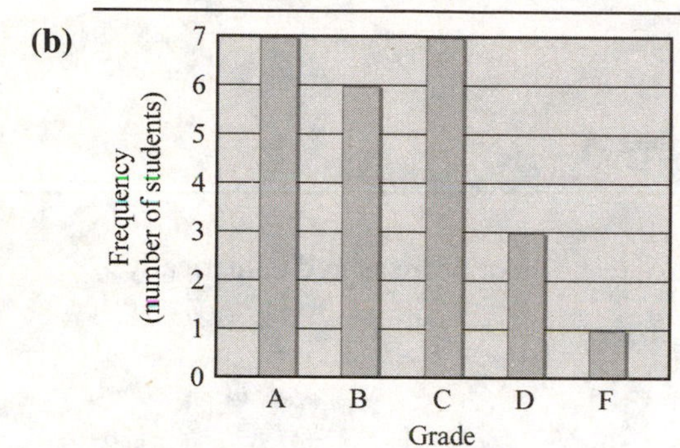

5.

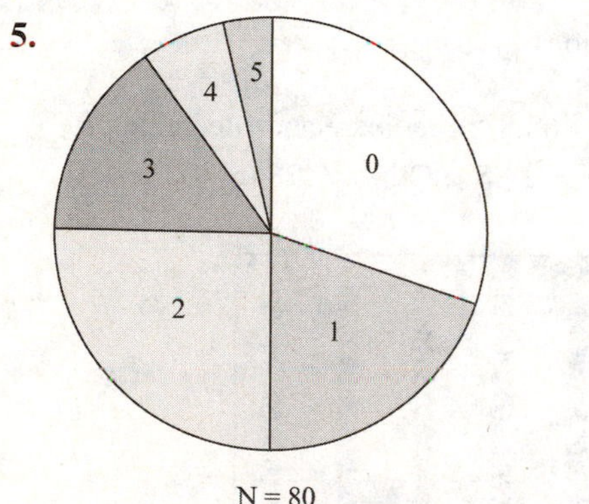

7.

Class Interval	Very close	Close	Nearby	Not too far	Far
Frequency	8	10	5	1	3

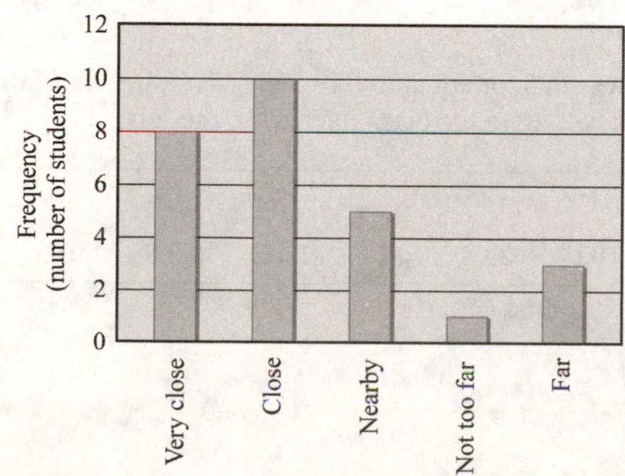

9.

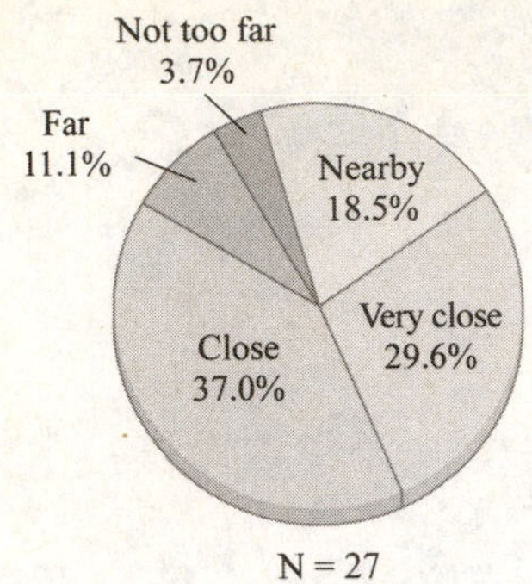

Slice "Very close": $\frac{8}{27}\times 360° \approx 106.7°$;

Slice "Close": $\frac{10}{27}\times 360° \approx 133.3°$;

Slice "Nearby": $\frac{5}{27}\times 360° \approx 66.7°$;

Slice "Not too far": $\frac{1}{27}\times 360° \approx 13.3°$;

Slice "Far": $\frac{3}{27}\times 360° = 40°$.

11. **(a)** $N = 4 + 6 + 7 + 5 + 6 + 8 + 2 + 2 = 40$

(b) 0%

(c) $\frac{5+6+8+2+2}{40} = 0.575 = 57.5\%$

13. **(a)** qualitative; it is a variable that describes characteristics that cannot be measured numerically.

(b) $0.47222 \times 19{,}548 \approx 9231$ died due to an accident

(c) Other: $0.16124 \times 360° \approx 58°$

15. **(a)** $0.228 \times \$2{,}900{,}000{,}000{,}000 = \$661{,}200{,}000{,}000$ (\$661.2 billion)

(b) Defense: $0.228 \times 360° \approx 82.1°$; Social Security: $0.215 \times 360° \approx 68.8°$; Medicare and Medicaid: $0.191 \times 360° \approx 77.4°$; Interest on the public debt: $0.082 \times 360° \approx 29.5°$; Other: $0.284 \times 360° \approx 102.2°$.

17.

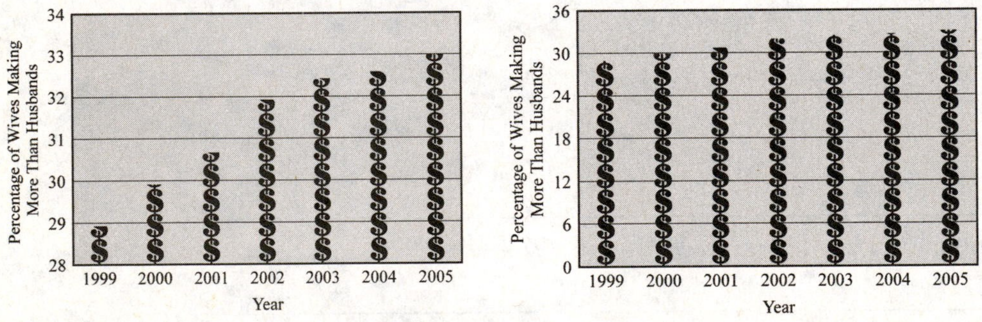

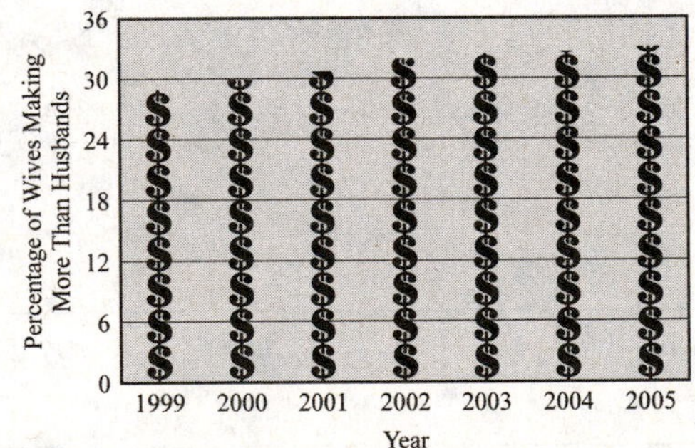

19. **(a)** $60 - 48 = 12$ ounces

(b) The third class interval: "more than 72 ounces and less than or equal to 84 ounces." Values that fall exactly on the boundary between two class intervals belong to the class interval to the left.

(c)

Frequency	15	24	41	67	119	184	142	26	5	2
Percent	2.4	3.8	6.6	10.7	19.0	29.4	22.7	4.2	0.8	0.3

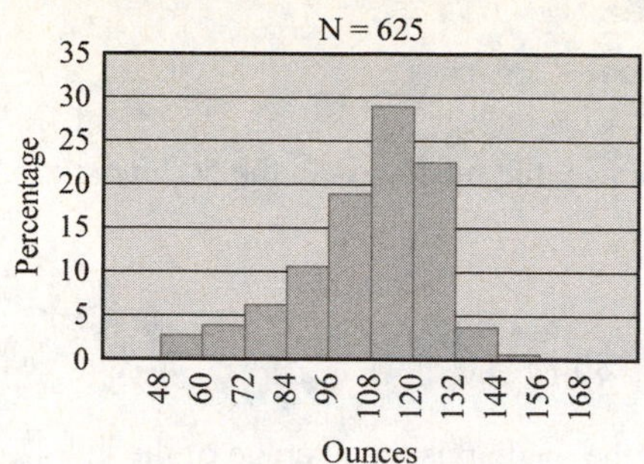

21. (a) $8 + 1 = 9$ (use Figure 14-18(b))

(b) $4 - 1 = 3$

B. Means and Medians

23. (a) average $= \dfrac{3-5+7+4+8+2+8-3-6}{9} = 2$

(b) The ordered data set is {-6, -5, -3, 2, 3, 4, 7, 8, 8}. The locator of the 50th percentile is $L = (0.50)(9) = 4.5$. Since L is not a whole number, the 50th percentile is located in the 5th position in the list. Hence, the median is 3.

(c) The average of the new data set is most easily found by computing $\dfrac{(9\times 2)+2}{10} = 2$. The new ordered data set is {-6, -5, -3, 2, 2, 3, 4, 7, 8, 8}. The locator of the 50th percentile is $L = (0.50)(10) = 5$. Since L is a whole number, the 50th percentile is the average of the 5th and 6th numbers (2 and 3) in the ordered list. So, the median of this new data set is 2.5.

25. (a) average $= \dfrac{0+1+2+3+4+5+6+7+8+9}{10} = \dfrac{45}{10} = 4.5$

Since the data set is already ordered, the locator of the median is $L = (0.50)(10) = 5$. Since L is a whole number, the median is the average of the 5th and 6th numbers (4 and 5) in the data set. The median is 4.5.

(b) average $= \dfrac{1+2+3+4+5+6+7+8+9}{9} = \dfrac{45}{9} = 5$

The locator of the median is $L = (0.50)(9) = 4.5$. Since L is not a whole number, the median is the 5th number in the data set. Hence the median is 5.

(c) average $= \dfrac{1+2+3+4+5+6+7+8+9+10}{10} = \dfrac{55}{10} = 5.5$

The locator of the median is $L = (0.50)(10) = 5$. Since L is a whole number, the median is the average of the 5th and 6th numbers (5 and 6) in the data set. The median is 5.5.

(d) average $= \dfrac{a+2a+3a+4a+5a+6a+7a+8a+9a+10a}{10} = \dfrac{55a}{10} = 5.5a$

The locator of the median is $L = (0.50)(10) = 5$. Since L is a whole number, the median is the average of the 5th and 6th numbers ($5a$ and $6a$) in the data set. So, the median is $5.5a$. (see (c)).

27. **(a)** average $= \dfrac{1+2+3+4+5+\ldots+98+99}{99} = \dfrac{\frac{99\times 100}{2}}{99} = 50$

(b) The locator for the median is $L = (0.50)(99) = 49.5$. Hence, the median is in the 50th position. That is, the median is 50.

29. **(a)** average $= \dfrac{24\times 0+16\times 1+20\times 2+12\times 3+5\times 4+3\times 5}{24+16+20+12+5+3} = \dfrac{127}{80} = 1.5875$

(b) The locator for the median is $L = (0.50)(80) = 40$. So, the median is the average of the 40th and 41st scores. The 40th score is 1, and the 41st score is 2. Thus, the median is 1.5.

31. **(a)** Since the number of quiz scores N is not given, we can compute the average as

$$\frac{(0.07N)\times 4+(0.11N)\times 5+(0.19N)\times 6+(0.24N)\times 7+(0.39N)\times 8}{N}$$

$$= 0.07\times 4+0.11\times 5+0.19\times 6+0.24\times 7+0.39\times 8 = 6.77$$

(As the calculation shows, the number of scores is not important. If one likes, they can assume that there were 100 scores.)

(b) Since 50% of the scores were at 7 or below, the median is 7.

C. Percentiles and Quartiles

33. The ordered data set is {-6, -5, -3, 2, 3, 4, 7, 8, 8}.

(a) The locator of the 25th percentile is $L = (0.25)(9) = 2.25$. By rounding up, we find that the first quartile is the 3rd number in the ordered list. That is, $Q_1 = -3$.

(b) The locator of the 75th percentile is $L = (0.75)(9) = 6.75$. By rounding up, we find that the third quartile is the 7th number in the ordered list. That is, $Q_3 = 7$.

(c) The new ordered data set is {-6, -5, -3, 2, 2, 3, 4, 7, 8, 8}. The locator of the 25th percentile is $L = (0.25)(10) = 2.5$. Rounding up, we find that the first quartile is the 3rd number in the ordered list. That is, $Q_1 = -3$. The locator of the 75th percentile is $L = (0.75)(10) = 7.5$. Rounding up, we find that the third quartile is the 8th number in the ordered list. That is, $Q_3 = 7$.

35. **(a)** Since the data set is already ordered, the locator of the 75th percentile is given by L = (0.75)(100) = 75. Since this is a whole number, the 75th percentile is the average of the 75th and 76th numbers in the list. That is, the 75th percentile is 75.5.

The locator of the 90th percentile is given by $L = (0.90)(100) = 90$. Since this is a whole number, the 90th percentile is the average of the 90th and 91st numbers in the list. That is, the 90th percentile is 90.5.

(b) The locator of the 75th percentile is given by $L = (0.75)(101) = 75.75$. Since this not is a whole number, the 75th percentile is the 76th number in the list. That is, the 75th percentile is 75.

The locator of the 90th percentile is given by $L = (0.90)(101) = 90.9$. Since this is not a whole number, the 90th percentile is the 91st number in the list. That is, the 90th percentile is 90.

(c) The locator of the 75th percentile is given by $L = (0.75)(99) = 74.25$. Since this not is a whole number, the 75th percentile is the 75th number in the list. That is, the 75th percentile is 75.

The locator of the 90th percentile is given by $L = (0.90)(99) = 89.1$. Since this is not a whole number, the 90th percentile is the 90th number in the list. That is, the 90th percentile is 90.

(d) The locator of the 75th percentile is given by $L = (0.75)(98) = 73.5$. Since this not is a whole number, the 75th percentile is the 74th number in the list. That is, the 75th percentile is 74.

The locator of the 90th percentile is given by $L = (0.90)(98) = 88.2$. Since this is not a whole number, the 90th percentile is the 89th number in the list. That is, the 90th percentile is 89.

37. **(a)** The Cleansburg Fire Department consists of 2 + 7 + 6 + 9 + 15 + 12 + 9 + 9 + 6 + 4 = 79 firemen. The locator of the first quartile is thus given by L = (0.25)(79) = 19.75. So, the first quartile is the 20th number in the ordered data set. That is, $Q_1 = 29$.

(b) The locator of the third quartile is given by L = (0.75)(79) = 59.25. So, the third quartile is the 60th number in the ordered data set. That is, $Q_3 = 32$.

(c) The locator of the 90th percentile is given by L = (0.90)(79) = 71.1. So, the 90th percentile is the 72nd number in the ordered data set or 37.

39. **(a)** The 747,266th score, $d_{747,266}$. [The locator is given by $L = 747{,}265.5$.]

(b) The 373,633rd score, $d_{373,633}$. [The locator is given by $L = 373{,}632.75$.]

(c) The 1,120,899th score, $d_{1,120,899}$. [The locator is given by $L = 1{,}120{,}898.25$.]

D. Box Plots and Five-Number Summaries

41. **(a)** $Min = -6,\ Q_1 = -3, M = 3, Q_3 = 7,\ Max = 8$

(b)

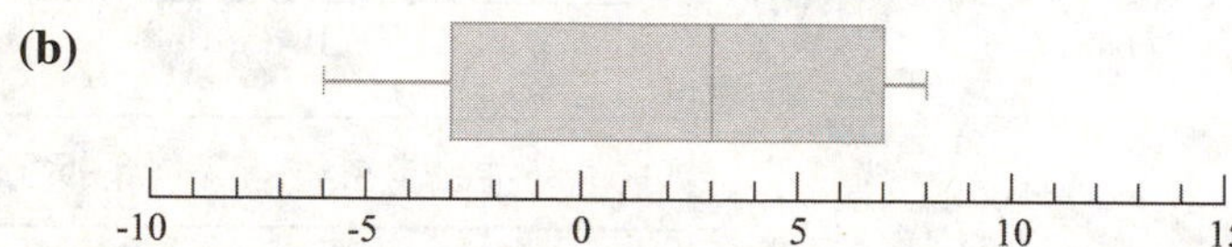

43. **(a)** $Min = 25,\ Q_1 = 29, M = 31, Q_3 = 32,\ Max = 39$

(b)

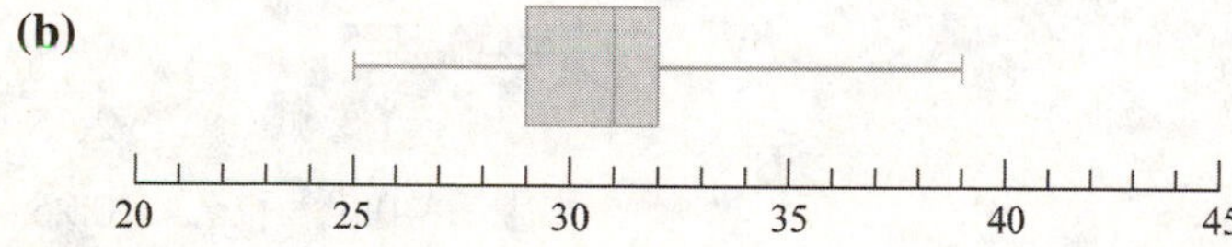

45. **(a)** Between \$33,000 and \$34,000 (corresponding the vertical line in the middle of the box)

(b) \$40,000

(c) The vertical line indicating the median salary in the engineering box plot is to the right of the box in the agriculture box plot.

E. Ranges and Interquartile Ranges

47. **(a)** 8 – (–6) = 14

(b) From exercise 33, $Q_1 = -3, Q_3 = 7$. $IQR = 7 - (-3) = 10$

49. **(a)** \$156,000 - \$115,000 = \$41,000

(b) At least 171 homes

51. **(a)** Note that $1.5 \times IQR = 1.5 \times 3 = 4.5$. Any number bigger than $Q_3 + 1.5 \times IQR = 12 + 4.5 = 16.5$ is an outlier.

(b) Any number smaller than $Q_1 - 1.5 \times IQR = 9 - 4.5 = 4.5$ is also an outlier.

(c) Since 1 is the only number smaller than 4.5 and 24 is the only number bigger than 16.5, the numbers 1 and 24 are the only outliers in this data set.

53. The *IQR* for 18-year old U.S. males is 71 – 67 = 4 inches. Since $1.5 \times IQR = 1.5 \times 4 = 6$, an outlier is any height more than $Q_3 + 1.5 \times IQR = 71 + 6 = 77$ inches or any height less than 67 - 6 = 61 inches.

F. Standard Deviations

55. **(a)** $A = 5$, so $x - A = 0$ for every number x in the data set. The standard deviation is 0.

(b) $A = \dfrac{0+5+5+10}{4} = 5$

x	$x - 5$	$(x-5)^2$
0	–5	25
5	0	0
5	0	0
10	5	25
		50

Standard deviation $= \sqrt{\dfrac{50}{4}} = \dfrac{5\sqrt{2}}{2} \approx 3.5$

(c) $A = \dfrac{0+10+10+20}{4} = 10$

x	$x - 10$	$(x-10)^2$
0	–10	100
10	0	0
10	0	0
20	10	100
		200

Standard deviation $= \sqrt{\dfrac{200}{4}} = 5\sqrt{2} \approx 7.1$

57. **(a)** $A = \dfrac{0+1+2+3+4+5+6+7+8+9}{10} = 4.5$

x	$x - 4.5$	$(x-4.5)^2$
0	–4.5	20.25
1	–3.5	12.25
2	–2.5	6.25
3	–1.5	2.25
4	–0.5	0.25
5	0.5	0.25
6	1.5	2.25
7	2.5	6.25
8	3.5	12.25
9	4.5	20.25
		82.5

Standard deviation $= \sqrt{\dfrac{82.5}{10}} \approx 2.87$

(b) $A = \dfrac{1+2+3+4+5+6+7+8+9+10}{10} = 5.5$

x	$x - 5.5$	$(x-5.5)^2$
1	–4.5	20.25
2	–3.5	12.25
3	–2.5	6.25
4	–1.5	2.25
5	–0.5	0.25
6	0.5	0.25
7	1.5	2.25
8	2.5	6.25
9	3.5	12.25
10	4.5	20.25
		82.5

Standard deviation $= \sqrt{\dfrac{82.5}{10}} \approx 2.87$

Note that each data point is simply located one unit to the right of that data set given in (a). So, the spread of the data set has not changed.

(c) $A = \frac{6+7+8+...+14+15}{10} = 10.5$

x	$x - 10.5$	$(x-10.5)^2$
6	–4.5	20.25
7	–3.5	12.25
8	–2.5	6.25
9	–1.5	2.25
10	–0.5	0.25
11	0.5	0.25
12	1.5	2.25
13	2.5	6.25
14	3.5	12.25
15	4.5	20.25
		82.5

Standard deviation = $\sqrt{\frac{82.5}{10}} \approx 2.87$

Note again that each data point is 6 units to the right of where it appeared in (a). So, the mean has changed (it is 6 units bigger), but the spread of the data has not. So, the standard deviation is the same as in (a).

G. Miscellaneous

59. 0 (frequency 24)

61. 4, 5, and 8 (frequency 5)

63. Caucasian (has largest percent or frequency)

JOGGING

65. Let x = score Mike needs on the next exam.

$$\frac{5 \cdot 88 + x}{6} = 90$$
$$5 \cdot 88 + x = 540$$
$$x = 100$$

67. Since the median is the average of $d_{732,872}$ and $d_{732,873}$, the locator of the median is L = 732,872. Since $(0.5)(N) = L$ = 732,872, we find N = 1,465,744.

69. **(a)** {1, 1, 1, 1, 6, 6, 6, 6, 6, 6} Average = 4; Median = 6.

(b) {1, 1, 1, 1, 1, 1, 6, 6, 6, 6} Average = 3; Median =1.

(c) {1, 1, 6, 6, 6, 6, 6, 6, 6, 6} Average = 5; $Q_1 = 6$.

(d) {1, 1, 1, 1, 1, 1, 1, 1, 6, 6} Average = 2; $Q_3 = 1$.

71. From histogram (a) one can deduce that the median team salary is between \$70 and \$100 million. From histogram (b) one can deduce that the median team salary is between \$50 and \$80 million. It follows that the median team salary must be between \$70 and \$80 million.

73. **(a)** 4

$$\frac{\text{Column area over interval } 30-35}{\text{Column area over interval } 20-30} = \frac{5 \times h}{10 \times 1} = \frac{50\%}{25\%}$$

and so $5h = 20$, $h = 4$.

(b) 0.4

$$\frac{\text{Column area over interval } 35-45}{\text{Column area over interval } 20-30} = \frac{10 \times h}{10 \times 1} = \frac{10\%}{25\%}$$

and so $h = 0.4$.

(c) 0.4

$$\frac{\text{Column area over interval } 45-60}{\text{Column area over interval } 20-30} = \frac{15 \times h}{10 \times 1} = \frac{15\%}{25\%}$$

and so $h = 0.4$.

75. **(a)** Male: 10%, Female: 20%

(b) Male: 80%, Female: 90%

(c) The figures for both schools were combined. A total of 820 males were admitted out of a total of 1200 that applied–an admission rate for males of approximately 68.3%. Similarly, a total of 460 females were admitted out of a total of 900 that applied–an admission rate for females of approximately 51.1%.

(d) In this example, females have a higher percentage $\left(\frac{100}{500} = 20\%\right)$ than males $\left(\frac{20}{200} = 10\%\right)$ for admissions to the School of Architecture and also a higher percentage $\left(\frac{360}{400} = 90\%\right)$ than males $\left(\frac{800}{1000} = 80\%\right)$ for the School of

Engineering. When the numbers are combined, however, females have a lower percentage $\left(\frac{100+360}{500+400} \approx 51.1\%\right)$ than males $\left(\frac{20+800}{200+1000} \approx 68.3\%\right)$ in total admissions. The reason that this apparent paradox can occur is purely a matter of arithmetic: Just because $\frac{a_1}{a_2} > \frac{b_1}{b_2}$ and $\frac{c_1}{c_2} > \frac{d_1}{d_2}$ it does not necessarily follow that $\frac{a_1+c_1}{a_2+c_2} > \frac{b_1+d_1}{b_2+d_2}$. The majority of males applied to the engineering school, which has a much higher acceptance rate than the School of Architecture. This, combined with the fact that more females applied to the School of Architecture than the School of Engineering, is why, overall, the percentage of males admitted is higher than the percentage of females admitted.

77. (a) $$\frac{(x_1+c)+(x_2+c)+(x_3+c)+\ldots+(x_N+c)}{N}$$
$$=\frac{(x_1+x_2+x_3+\ldots+x_N)+cN}{N}$$
$$=\frac{(x_1+x_2+x_3+\ldots+x_N)}{N}+\frac{cN}{N}=A+c$$

(b) Adding c to each value of a data set adds c to the average as well. Hence, subtracting the average A from each value of a data set (adding $-A$ to each value) subtracts A from the average creating a new average of 0.

79. (a) Let M be the maximum value of $\{x_1, x_2, x_3, \ldots, x_N\}$. The maximum value of $\{x_1+c, x_2+c, x_3+c, \ldots, x_N+c\}$ is $M + c$. Similarly, if the minimum value of $\{x_1, x_2, x_3, \ldots, x_N\}$ is m, then the minimum value of $\{x_1+c, x_2+c, x_3+c, \ldots, x_N+c\}$ is $m + c$. So, the range of $\{x_1+c, x_2+c, x_3+c, \ldots, x_N+c\}$ is $M + c - (m + c) = M - m$. That is, the same as the range of $\{x_1, x_2, x_3, \ldots, x_N\}$.

(b) Adding c to each value of a data set does not change how spread out the data set is. Since the standard deviation is a measure of spread, the standard deviation does not change. Specifically, the average of the data set will increase by c, but the deviations from the mean (and hence the standard deviation) will remain the same.

Chapter 15

WALKING

A. Random Experiments and Sample Spaces

1. (a) {*HHH, HHT, HTH, THH, TTH, THT, HTT, TTT*}
Note that there are $2^3 = 8$ outcomes consisting of all possible sequences of *H*'s and *T*'s of length three.

(b) {*SSS, SSF, SFS, FSS, FFS, FSF, SFF, FFF*}
Note the correspondence between the outcomes in (a) and (b).

(c) {0, 1, 2, 3}

(d) {0, 1, 2, 3}

3. {*ABCD, ABDC, ACBD, ACDB, ADBC, ADCB, BACD, BADC, BCAD, BCDA, BDAC, BDCA, CABD, CADB, CBAD, CBDA, CDAB, CDBA, DABC, DACB, DBAC, DBCA, DCAB, DCBA*}
The sample space consists of $N = 4\times3\times2\times1 = 24$ outcomes each of which represents an ordering of the letters *A*, *B*, *C*, and *D* (4 choices for first name chosen, 3 choices for second name chosen, etc.).

5. Answers will vary. A typical outcome is a string of 10 letters each of which can be either an *H* or a *T*. An answer like {*HHHHHHHHHH*, …,*TTTTTTTTTT*} is not sufficiently descriptive. An answer like{… *HTTHHHTHTH*, …, *TTHTHHTTHT*, …, *HHHTHTTHHT*, …} is better. An answer like $\{X_1X_2X_3X_4X_5X_6X_7X_8X_9X_{10}$: each X_i is either *H* or *T*} is best. Note: This sample space consists of $N = 2^{10} = 1024$ outcomes (2 possible outcomes at each of 10 stages of the experiment).

7. Answers will vary. An outcome is an ordered sequence of four numbers, each of which is an integer between 1 and 6 inclusive. The best answer would be something like $\{(n_1, n_2, n_3, n_4)$: each n_i is 1, 2, 3, 4, 5, or 6$\}$. An answer such as {(1,1,1,1), …, (1,1,1,6), …,(1,2,3,4),…, (3,2,6,2), …, (6,6,6,6)} showing a few typical outcomes is possible, but not as good. An answer like {(1,1,1,1),…, (2,2,2,2), …, (6,6,6,6)} is not descriptive enough. Note: This sample space consists of $N = 6^4 = 1296$ outcomes (6 possible outcomes at each of 4 stages of the experiment).

B. The Multiplication Rule

9. (a) $9\times26^3\times10^3 = 158{,}184{,}000$

(b) $1\times26^3\times10^2\times1 = 1{,}757{,}600$

(c) $9\times26\times25\times24\times9\times8\times7 = 70{,}761{,}600$

11. (a) If Dolores chooses to wear high heels, she could also wear either a formal dress or jeans and a blouse. There are $2\times(4+3\times4) = 32$ such combinations. If Dolores chooses to wear tennis shoes, she could also wear jeans and a blouse. There are $2\times3\times5 = 30$ such combinations. So, Dolores has 32 + 30 = 62 possible outfits.

(b) There are $2\times3\times5 = 30$ outfits that could be made in which Dolores is wearing a t-shirt. So the sweater adds 30 more possibilities to the options in (a). That is, she could make 62 + 30 = 92 outfits.

13. (a) 8! = 40,320

(b) 40,320 – 1 = 40,319 (there is only one way in which all of the books are in order)

15. (a) $35\times34\times33 = 39{,}270$

(b) $15\times14\times13=2730$.

(c) The total number of all-female committees is $15\times14\times13=2730$.
The total number of all-male committees is $20\times19\times18=6840$.
So, there are 2730 + 6840 = 9570 committees of the same gender.

(d) The remaining $35\times34\times33-(15\times14\times13+20\times19\times18)=39,270-(2730+6840)=$ 29,700 committees are mixed.

17. (a) $9\times10^5\times5=4,500,000$
There are 9 choices for the first digit (1-9), 10 choices for the next 5 digits (0-9), and 5 digits for the last digit (0, 2, 4, 6, 8).

(b) $9\times10^5\times2=1,800,000$
There are 2 choices for the last digit (either a 0 or a 5).

(c) $9\times10^4\times4=360,000$
The last 2 digits must be 00, 25, 50 or 75.

C. Permutations and Combinations

19. (a) ${}_{10}P_2=10\times9=90$

(b) ${}_{10}C_2=\dfrac{10\times9}{2\times1}=\dfrac{90}{2}=45$

(c) ${}_{10}P_3=10\times9\times8=720$

(d) ${}_{10}C_3=\dfrac{720}{3!}=\dfrac{720}{6}=120$

21. (a) ${}_{10}C_9=\dfrac{10!}{(10-9)!9!}=\dfrac{10!}{1!9!}=10$

(b) ${}_{10}C_8=\dfrac{10!}{(10-8)!8!}=\dfrac{10!}{2!8!}=\dfrac{10\times9}{2\times1}=45$

(c) ${}_{200}C_{199}=\dfrac{200!}{(200-199)!\times199!}=\dfrac{200!}{1!199!}=200$

(d) ${}_{200}C_{198}=\dfrac{200!}{(200-198)!198!}=\dfrac{200!}{2!198!}=\dfrac{200\times199}{2\times1}=19,900$

23. (a) ${}_{11}C_4=\dfrac{11!}{(11-4)!4!}=\dfrac{11!}{7!4!}=330$

(b) ${}_{11}C_7=\dfrac{11!}{(11-7)!7!}=\dfrac{11!}{4!7!}=330$

(c) ${}_{50}C_3=\dfrac{50!}{(50-3)!3!}=\dfrac{50!}{47!3!}=\dfrac{50\times49\times48}{3\times2\times1}=19,600$

(d) ${}_{50}C_{47} = \dfrac{50!}{(50-47)!47!} = \dfrac{50!}{3!47!} = \dfrac{50\times49\times48}{3\times2\times1} = 19{,}600$

25. (a) ${}_{11}C_3 + {}_{11}C_4 = \dfrac{11!}{8!3!} + \dfrac{11!}{7!4!} = 165 + 330 = 495$

(b) ${}_{12}C_4 = \dfrac{12!}{8!4!} = 495$ (Note that this is the same as the answer in part (a)!)

(c) ${}_{12}C_4 + {}_{12}C_5 = \dfrac{12!}{8!4!} + \dfrac{12!}{7!5!} = 495 + 792 = 1287$

(d) ${}_{13}C_5 = \dfrac{13!}{8!5!} = 1287$ (Note that this is the same as the answer in part (c)! Do you spot a pattern?)

27. First, note that ${}_{150}P_{50} = \dfrac{150!}{100!} = \dfrac{150\times149\times148\times\cdots\times3\times2\times1}{100\times99\times98\times\cdots\times3\times2\times1}$ and ${}_{150}P_{51} = \dfrac{150!}{99!} = \dfrac{150\times149\times148\times\cdots\times3\times2\times1}{99\times98\times97\times\cdots\times3\times2\times1}$.
It follows that ${}_{150}P_{51} = 100\times{}_{150}P_{50}$. That is, ${}_{150}P_{51}$ is 100 times bigger than ${}_{150}P_{50}$. Thus, ${}_{150}P_{51} = 6.12\times10^{104}\times100 = 6.12\times10^{106}$.

29. ${}_{150}C_{50} = \dfrac{150!}{100!50!} = \dfrac{150\times149\times148\times\cdots\times3\times2\times1}{(100\times99\times98\times\cdots\times2\times1)(50\times49\times48\times\cdots\times2\times1)}$ and

${}_{150}C_{51} = \dfrac{150!}{99!51!} = \dfrac{150\times149\times148\times\cdots\times3\times2\times1}{(99\times98\times97\times\cdots\times2\times1)(51\times50\times\cdots\times2\times1)}$. It follows that $51\times({}_{150}C_{51}) = 100\times({}_{150}C_{50})$.

That is, ${}_{150}C_{51}$ is $\dfrac{100}{51}$ times bigger than ${}_{150}C_{50}$. Thus, ${}_{150}C_{51} = \dfrac{100}{51}\times2.01\times10^{40} \approx 3.94\times10^{40}$.

31. (a) 60,949,324,800

(b) 167,960

(c) 167,960

33. (a) ${}_{15}P_4$; the order the candidates are selected to fill these openings matters (i.e., this is an ordered selection).

(b) ${}_{15}C_4$; the order in which the delegation is selected is irrelevant.

35. (a) ${}_{30}C_{18}$; the order the songs are selected for the set list does not matter (i.e., this an unordered selection).

(b) ${}_{30}P_{18}$; the order the songs are selected for the CD (sequencing) matters.

D. General Probability Spaces

37. (a) $\Pr(o_1)+\Pr(o_2)+\Pr(o_3)+\Pr(o_4)+\Pr(o_5)=1$

$$0.22+0.24+3\Pr(o_3)=1$$

$$\Pr(o_3)=0.18$$

The probability assignment is $\Pr(o_1)=0.22$, $\Pr(o_2)=0.24$, $\Pr(o_3)=0.18$, $\Pr(o_4)=0.18$, $\Pr(o_5)=0.18$.

(b) $\Pr(o_1)+\Pr(o_2)+\Pr(o_3)+\Pr(o_4)+\Pr(o_5)=1$

$0.22+0.24+[\Pr(o_4)+0.1]+\Pr(o_4)+0.1=1$

$\Pr(o_4)=0.17$

The probability assignment is $\Pr(o_1)=0.22$, $\Pr(o_2)=0.24$, $\Pr(o_3)=0.27$, $\Pr(o_4)=0.17$, $\Pr(o_5)=0.1$.

39. We know that $\Pr(A)=\frac{1}{5}$. If $\Pr(B)=x$, then $\Pr(C)=2x$ and $\Pr(D)=3x$. So, since these probabilities must sum to 1, we have $1/5 + x + 2x + 3x = 1$ which means that $6x = 4/5$ or $x = 2/15$. It follows that the probability assignment is $\Pr(A)=\frac{1}{5}, \Pr(B)=\frac{2}{15}, \Pr(C)=\frac{4}{15}, \Pr(D)=\frac{6}{15}$.

41. $\Pr(\text{blue}) = \Pr(\text{white}) = \frac{72°}{360°} = 0.2$; $\Pr(\text{green}) = \Pr(\text{yellow}) = \frac{54°}{360°} = 0.15$. The probability assignment is Pr(red) = 0.3, Pr(blue) = Pr(white) = 0.2, Pr(green) = Pr(yellow) = 0.15.

E. Events

43. **(a)** $E_1 = \{HHT, HTH, THH\}$

(b) $E_2 = \{HHH, TTT\}$

(c) $E_3 = \{\}$

(d) $E_4 = \{TTH, TTT\}$

45. **(a)** $E_1 = \{(1,1),(2,2),(3,3),(4,4),(5,5),(6,6)\}$

(b) $E_2 = \{(1,1),(1,2),(2,1),(6,6)\}$

(c) $E_3 = \{(1,6),(2,5),(3,4),(4,3),(5,2),(6,1),(5,6),(6,5)\}$

47. **(a)** $E_1 = \{HHHHHHHHHH\}$

(b) $E_2 = \{HHHHHHHHHT, HHHHHHHHTH, HHHHHHHTHH, HHHHHHTHHH, HHHHHTHHHH,$
$HHHHTHHHHH, HHHTHHHHHH, HHTHHHHHHH, HTHHHHHHHH, THHHHHHHHH\}$

(c) $E_3 = \{TTTTTTTTTH, TTTTTTTTHT, TTTTTTTHTT, TTTTTTHTTT, TTTTTHTTTT,$
$TTTTHTTTTT, TTTHTTTTTT, TTHTTTTTTT, THTTTTTTTT, HTTTTTTTTT, TTTTTTTTTT\}$

F. Equiprobable Spaces

49. **(a)** $\Pr(E_1)=\frac{3}{8}=0.375$; the sample space has 8 equally likely outcomes (see Exercise 1(a)). Of these outcomes, three of them constitute the event $E_1 = \{HHT, HTH, THH\}$ (see Exercise 43(a)).

(b) $\Pr(E_2)=\frac{2}{8}=0.25$; see Exercises 1(a) and 43(b).

(c) $\Pr(E_3)=0$; see Exercises 1(a) and 43(c).

(d) $\Pr(E_4) = \frac{2}{8} = 0.25$; see Exercises 1(a) and 43(d).

51. (a) $\Pr(E_1) = \frac{6}{36} = \frac{1}{6}$; the sample space has $6^2 = 36$ equally likely outcomes. Of these outcomes, six of them constitute the event $E_1 = \{(1,1),(2,2),(3,3),(4,4),(5,5),(6,6)\}$ (see Exercise 45(a)).

(b) $\Pr(E_2) = \frac{4}{36} = \frac{1}{9}$ (see Exercise 45(b))

(c) $\Pr(E_3) = \frac{8}{36} = \frac{2}{9}$ (see Exercise 45(c))

53. (a) $\Pr(E_1) = \frac{1}{1024} \approx 0.001$; the sample space has $2^{10} = 1024$ equally likely outcomes (see Exercise 5). Of these outcomes, only one of them constitutes the event that none of the tosses come up tails: $E_1 = \{HHHHHHHHHH\}$ (see Exercise 47(a)).

(b) $\Pr(E_2) = \frac{10}{1024} = \frac{5}{512} \approx 0.01$; (see Exercise 47(b))

(c) $\Pr(E_3) = \frac{11}{1024} \approx 0.01$; (see Exercise 47(c))

55. The total number of outcomes in this random experiment is $2^{10} = 1024$.

(a) There is only one way to get all ten correct. So, $\Pr(\text{getting 10 points}) = \frac{1}{1024}$. If it helps, think of this as being similar to Exercise 53(a). If the "key" to the quiz is all "True" (or "Heads"), there is only one outcome in which the student earns exactly 10 points.

(b) There is only one way to get all ten incorrect (and hence a score of –5). So, $\Pr(\text{getting -5 points}) = \frac{1}{1024}$. Again, you can think of this as the "key" to the quiz consisting of all "True" answers. In this case, there is only one outcome in which the student earns a score of -5 (their quiz would have all "False" answers. See Exercise 53 and use True=Heads and False=Tails.

(c) In order to get 8.5 points, the student must get exactly 9 correct answers and 1 incorrect answer. There are ${}_{10}C_1 = 10$ ways to select which answer would be answered incorrectly. So, $\Pr(\text{getting 8.5 points}) = \frac{10}{1024} = \frac{5}{512}$. See also Exercise 53(b).

(d) In order to get 8 or more points, the student must get at least 9 correct answers (if they only get 8 correct answers, they lose a point for guessing two incorrect answers and score 7 points). So, $\Pr(\text{getting 8 or more points}) = \Pr(\text{getting 8.5 points}) + \Pr(\text{10 points}) = \frac{10}{1024} + \frac{1}{1024} = \frac{11}{1024}$. See also Exercise 53(c).

(e) If the student gets 6 answers correct, they score $6 - 4 \times 0.5 = 4$ points. If the student gets 7 answers correct, they score $7 - 3 \times 0.5 = 5.5$ points. So, there is no chance of getting exactly 5 points. Hence, $\Pr(\text{getting 5 points}) = 0$.

(f) If the student gets 8 answers correct, they score $8-2\times0.5=7$ points. There are ${}_{10}C_2=45$ ways to select which 8 answers would be answered correctly. So, Pr(getting 7 points) $=\frac{45}{1024}$. In order to get 7 or more points, the student needs to answer at least 8 answers correctly. Pr(getting 7 or more points) =

Pr(getting 7 points) + Pr(getting 8.5 points) + Pr(10 points) $=\frac{45}{1024}+\frac{10}{1024}+\frac{1}{1024}=\frac{56}{1024}=\frac{7}{128}$.

57. (a) There are ${}_{15}C_4=1365$ ways to choose four delegates. If Alice is selected, there are ${}_{14}C_3=364$ ways to choose the other three delegates. So, Pr(Alice selected) $=\frac{364}{1365}=\frac{4}{15}$.

(b) Pr(Alice is not selected) $=1-\frac{364}{1365}=\frac{1001}{1365}=\frac{11}{15}$.

(c) There are ${}_{15}C_4=1,365$ ways to select four members, but only one way to select Alice, Bert, Cathy, and Dale. Pr(Alice, Bert, Cathy, and Dale selected) $=\frac{1}{{}_{15}C_4}=\frac{1}{1,365}$.

G. Odds

59. (a) $a=4$, $b=7$, $b-a=7-4=3$. The odds in favor of E are 4 to 3. That is, for every 7 times the experiment is run, we should expect the event E to occur 4 times and to not occur 3 times.

(b) $a=6$, $b=10$, $b-a=10-6=4$. The odds in favor of E are 6 to 4, or 3 to 2.

61. (a) $\Pr(E)=\frac{3}{3+5}=\frac{3}{8}$

(b) $\Pr(E)=1-\frac{8}{8+15}=1-\frac{8}{23}=\frac{15}{23}$

JOGGING

63. There are 35 ways to select a chair and 34 ways to select a secretary. From the remaining 33 members of the ski club, there are ${}_{33}C_3=5456$ ways to select three at-large members. This makes a total of $35\times34\times5456=6,492,640$ ways to select a planning committee.

65. (a) The event that X wins in 5 games can be described by {*YXXXX, XYXXX, XXYXX, XXXYX*}.

(b) The event that the series is over in 5 games can be described by {*YXXXX, XYXXX, XXYXX, XXXYX, XYYYY, YXYYY, YYXYY, YYYXY*}.

(c) It may be useful to describe the events (1) the series goes six games with X winning and (2) the series goes six games with Y winning.

	X wins	*Y* wins
six game series in which…	*YYXXXX, YXYXXX, YXXYXX, YXXXYX, XYYXXX, XYXYXX, XYXXYX, XXYYXX, XXYXYX, XXXYYX*	*XXYYYY, XYXYYY, XYYXYY, XYYYXY, YXXYYY, YXYXYY, YXYYXY, YYXXYY, YYXYXY, YYYXXY*

The event "the series is over in game 6" can be described by {*YYXXXX, YXYXXX, YXXYXX, YXXXYX, XYYXXX, XYXYXX, XYXXYX, XXYYXX, XXYXYX, XXXYYX, XXYYYY, XYXYYY, XYYXYY, XYYYXY, YXXYYY, YXYXYY, YXYYXY, YYXXYY, YYXYXY, YYYXXY*}

(d) If X wins the series, there are ${}_6C_3 = 20$ ways that the series can end in 7 games (X is listed last in all the outcomes of this event and wins 3 of the other 6 games). Similarly, if X wins the series there are a total of ${}_5C_3 = 10$ ways that the series can end in 6 games, ${}_4C_3 = 4$ ways that the series can end in 5 games, and there is ${}_3C_3 = 1$ way the series can end in 4 games. That is, there are 20 + 10 + 4 + 1 = 35 ways the series can end with X winning. Since there are also 35 ways the series can end with Y winning, the sample space has 70 outcomes.

67. (a) 10! = 3,628,800

(b) A circle of 10 people can be broken to form a line in 10 different ways. So there are $\frac{3,628,800}{10} = 362,880$ ways to form a circle.

(c) There are 2 choices as to which sex will start the line. Then, there are 5! ways to order the boys in the line and 5! ways to order the girls in the line. So, there are $2 \times 5! \times 5! = 28,800$ ways to form such a line.

(d) $\frac{2 \times 5! \times 5!}{10} = 2,880$ ways

69. Suppose, for the moment, that the order the teams (but not their members) were selected mattered. There are $({}_{15}C_5) \cdot ({}_{10}C_5) \cdot ({}_5C_5) = 756,756$ ways to select the teams. Since the order that the teams are selected does not matter, and there are 3! = 6 ways to rearrange the teams, there are actually $\frac{756,756}{6} = 126,126$ ways to select the teams.

As an alternative method of solution, think in the following way: Start with an ordered list of the study group. The first person on the list must be in *some* group. Then, select 4 of the remaining 14 members to be in that group (${}_{14}C_4 = 1001$ ways to do this). The first person on the list of the remaining members must be in some group (one of the two groups left to be formed). Then, select 4 of the remaining 9 members to be in that group (${}_9C_4 = 126$ ways to do this). Finally, the remaining 5 people on the list must belong to the study group not yet formed. So, there are $1001 \times 126 = 126,126$ ways to form the three groups.

71. The total number of possible outcomes (strings of 20 H's and T's) in this experiment is $2^{20} = 1,048,576$.

(a) Number of ways to choose the positions of 10 H's: ${}_{20}C_{10} = 184,756$.

$$\Pr(10\ H\text{'s and } 10\ T\text{'s}) = \frac{184,756}{1,048,576} = \frac{46,189}{262,144} \approx 0.176$$

(b) Number of ways to choose the positions of 3 H's: ${}_{20}C_3 = 1140$.

$$\Pr(3\ H\text{'s and } 17\ T\text{'s}) = \frac{1140}{1,048,576} = \frac{285}{262,144} \approx 0.001$$

(c) Pr(3 or more H's) = 1 – Pr(0 H's) – Pr(1 H's) – Pr(2 H's)

$$=1-\frac{{}_{20}C_0}{2^{20}}-\frac{{}_{20}C_1}{2^{20}}-\frac{{}_{20}C_2}{2^{20}}$$

$$=1-\frac{1}{2^{20}}-\frac{20}{2^{20}}-\frac{190}{2^{20}}$$

$$=1-\frac{211}{1,048,576}$$

$$=\frac{1,048,365}{1,048,576}\approx 0.9998$$

73. The total number of 5-card draw poker hands is 2,598,960. The number of hands with all 5 cards the same color is $\frac{52\times 25\times 24\times 23\times 22}{5!}=131,560$. So, Pr(all 5 cards have same color) $=\frac{131,560}{2,598,960}=\frac{253}{4,998}\approx 0.05$.

75. The total number of 5-card draw poker hands is 2,598,960. The number of ways to get 10, J, Q, K, A of any suit (including all the same suit) is $\frac{20\times 16\times 12\times 8\times 4}{5!}=1024.$

There are 4 ways for these cards to all be the same suit. So, there are 1024 – 4 = 1020 ways to get an ace-high straight. Thus, Pr(ace-high straight) $=\frac{1020}{2,598,960}=\frac{1}{2548}\approx 0.00039$.

77. Pr(win Bet 1) = 1 – Pr(never roll a 6) = 1 – Pr(roll outcome other than a 6 four times) $= 1-\left(\frac{5}{6}\right)^4\approx 0.5177$.

Pr(win Bet 2) = 1 – Pr(never roll boxcars) $=1-\left(\frac{35}{36}\right)^{24}\approx 0.4914$. So, Bet 1 (that of rolling at least one 6 in four rolls of a die) is better.

79. The key word in this exercise is "at least." The sample space consists of all possible outcomes. There are ${}_{500}C_5$ ways in which 5 tickets can be selected from the 500 to be "winning" tickets. For the event that you win at least one prize to occur, it could happen that you win one prize, two prizes, or three prizes. So, in symbolic notation,

Pr(winning at least one prize) = Pr(win 1 prize) + Pr(win 2 prizes) + Pr(win 3 prizes)

$$=\frac{{}_5C_1\cdot{}_{495}C_2}{{}_{500}C_5}+\frac{{}_5C_2\cdot{}_{495}C_1}{{}_{500}C_5}+\frac{{}_5C_3\cdot{}_{495}C_0}{{}_{500}C_5}$$

Chapter 16

WALKING

A. Normal Curves

1. **(a)** $\mu = 83$ lb; it is located in the exact middle of the distribution.

(b) $M = 83$ lb; it too is located in the exact middle of the distribution.

(c) $\sigma = 90$ lb. – 83 lb. = 7 lb; it is the horizontal distance between the middle of the distribution and the inflection point *P*.

3. **(a)** $\frac{72 \text{ in.} + 78 \text{ in.}}{2} = 75$ in.

(b) Using the result from (a), we see that $\sigma = 78$ in. – 75 in. = 3 in.

(c) $Q_3 \approx 75 \text{ in.} + 0.675 \times 3 \text{ in.} \approx 77$ in.

(d) $Q_1 \approx 75 \text{ in.} - 0.675 \times 3 \text{ in.} \approx 73$ in.

5. **(a)** $Q_3 \approx 81.2 \text{ lb.} + 0.675 \times 12.4 \text{ lb.} \approx 89.6$ lb

(b) $Q_1 \approx 81.2 \text{ lb.} - 0.675 \times 12.4 \text{ lb.} \approx 72.8$ lb

7. **(a)** $\mu = \frac{432.5 \text{ points} + 567.5 \text{ points}}{2}$

$= 500$ points

(b) $567.5 \approx 500 + 0.675 \times \sigma$

$\sigma \approx 100$ points

9. $94.7 \approx 81.2 + 0.675 \times \sigma$

$\sigma \approx 20$ in.

11. $\mu \neq M$; in any normal distribution these two values must be the same.

13. In a normal distribution the first and third quartiles are the same distance from the mean. In this distribution, $\mu - Q_1 = 195 - 180 = 15$ and $Q_3 - \mu = 220 - 195 = 25$.

B. Standardizing Data

15. **(a)** $\frac{45 \text{ kg} - 30 \text{ kg}}{15 \text{ kg}} = \frac{15 \text{ kg}}{15 \text{ kg}} = 1$

(b) $\frac{0 \text{ kg} - 30 \text{ kg}}{15 \text{ kg}} = \frac{-30 \text{ kg}}{15 \text{ kg}} = -2$

(c) $\frac{54 \text{ kg} - 30 \text{ kg}}{15 \text{ kg}} = \frac{24 \text{ kg}}{15 \text{ kg}} = 1.6$

(d) $\frac{3 \text{ kg} - 30 \text{ kg}}{15 \text{ kg}} = \frac{-27 \text{ kg}}{15 \text{ kg}} = -1.8$

17. **(a)** In a normal distribution, the third quartile is about 0.675 standard deviations above the mean. That is, $Q_3 \approx \mu + (0.675)\sigma$. In this case, it means $391 \approx 310 + (0.675)\sigma$ so that the standard deviation is $\sigma \approx 120$. Hence, the standardized value of 490 points is approximately $\frac{490 - 310}{120} = 1.5$.

(b) $\frac{250 - 310}{120} \approx -0.5$

(c) $\frac{220 - 310}{120} = -0.75$

(d) $\frac{442 - 310}{120} = 1.1$

19. -0.675

21. **(a)** $\frac{x - 183.5}{31.2} = -1$

$x - 183.5 = -31.2$

$x = 152.3$ ft

(b) $\frac{x - 183.5}{31.2} = 0.5$

$x - 183.5 = 15.6$

$x = 199.1$ ft

(c) $\frac{x - 183.5}{31.2} = -2.3$

$x - 183.5 = -71.76$

$x = 111.74$ ft

(d) $\frac{x-183.5}{31.2}=0$

$x-183.5=0$

$x=183.5$ ft

23. $\frac{84-50}{\sigma}=2$

$34=2\sigma$

$\sigma=17$ lb.

25. $\frac{50-\mu}{15}=3$

$50-\mu=45$

$\mu=5$

27. $\frac{20-\mu}{\sigma}=-2; \frac{100-\mu}{\sigma}=3$

$20-\mu=-2\sigma; 100-\mu=3\sigma$

$\mu=20+2\sigma$, so $100-(20+2\sigma)=3\sigma$

$100-20-2\sigma=3\sigma$

$80=5\sigma$

$\sigma=16$

$\mu=20+2\times16=52$

C. The 68 – 95 – 99.7 Rule

29. P is one standard deviation above the mean, and P' is one standard deviation below the mean.

$\mu=\frac{50+60}{2}=55$

$\sigma=60-55=5$

31. (a) 98.8 is two standard deviations above the mean and 85.2 is two standard deviations below the mean. So,

$\mu=\frac{85.2+98.8}{2}=92$

$\sigma=\frac{1}{2}(98.8-92)=3.4$

(b) $Q_1=92-0.675\times3.4=89.705$

$Q_3=92+0.675\times3.4=94.295$

33. 73.25 is the first quartile and 86.75 is the third quartile.

$\mu=\frac{73.25+86.75}{2}=80$

$Q_3-\mu\approx86.75-80=6.75$

$0.675\sigma\approx6.75$

$\sigma\approx10$

35. Since 84% of the data lies above one standard deviation of the mean, $\mu-\sigma=50.2$ cm. Thus, $\mu=6.1+50.2=56.3$ cm.

37. (a) 95% of the data lies within two standard deviations of the mean. Hence, 2.5% of the data are not within two standard deviations on each side of the mean. So, 97.5% of the data fall below $\mu+2\sigma$, the point two standard deviations above the mean.

(b) 97.5% of the data lies below $\mu+2\sigma$. Since 68% of the data lies within one standard deviation of the mean, 32% of the data is not within one standard deviation of the mean. So, 68% + 16% = 84% of the data fall below $\mu+\sigma$. It follows that $97.5\%-84\%=13.5\%$ of the data falls between $\mu+\sigma$ and $\mu+2\sigma$.

39. 9.9 has a standardized value of $\frac{9.9-12.6}{4}=-0.675$. Also, 16.6 has a standardized value of $\frac{16.6-12.6}{4}=1$. So, 25% of the data is below the first quartile of 9.9. Also, 84% of the data is below 16.6. So, 84% - 25% = 59% of the data is between 9.9 and 16.6.

D. Approximately Normal Data Sets

41. (a) 52 points

(b) 50%

(c) $\frac{41-52}{11}=-1, \frac{63-52}{11}=1$

So, since about 68% of data fall within one standard deviation of the mean (between standardized scores of -1 and 1) in a normal distribution, we would estimate that 68% of the students would score between 41 and 63 points.

(d) $\frac{1}{2}(100\% - 68\%) = 16\%$

43. (a) $Q_1 \approx 52 - 0.675 \times 11 \approx 44.6$ points

(b) $Q_3 \approx 52 + 0.675 \times 11 \approx 59.4$ points

45. (a) $\frac{99-125}{13} = -2, \frac{151-125}{13} = 2$
95% have blood pressure between 99 and 151 mm (i.e., between standardized scores of -2 and 2). 95% of 2000 patients is 1900 patients ($0.95 \times 2000 = 1900$).

(b) 112 is one standard deviation below the mean. 151 is two standard deviations above the mean. The percentage of patients falling between one standard deviation below the mean and two standard deviations above the mean is 68% + 13.5% = 81.5%. 81.5% of the 2000 patients is 1630 patients.

47. (a) $\frac{112-125}{13} = -1$
the 16th percentile

(b) $\frac{138-125}{13} = 1$
the 84th percentile

(c) $\frac{164-125}{13} = 3$
the 99.85th percentile

49. (a) $\frac{11-12}{0.5} = -2, \frac{13-12}{0.5} = 2$
95%

(b) $\frac{12-12}{0.5} = 0, \frac{13-12}{0.5} = 2$
$\frac{1}{2}(95\%) = 47.5\%$

(c) Because the chance of the bag weighing between 11 and 12 ounces is the same as the chance that it weighs between 12 and 13 ounces, we can use our answer to part (b) and symmetry to obtain an answer of 47.5% + 50% = 97.5%.

51. (a) $\frac{11-12}{0.5} = -2$
$\frac{1}{2}(100\% - 95\%) = 2.5\%$
$0.025 \times 500 = 12.5 \approx 13$ bags

(b) $\frac{11.5-12}{0.5} = -1$
$\frac{1}{2}(100\% - 68\%) = 16\%$
$0.16 \times 500 = 80$ bags

(c) 50%
$0.50 \times 500 = 250$ bags

(d) $\frac{12.5-12}{0.5} = 1$
$\frac{1}{2}(68\%) + 50\% = 84\%$
$0.84 \times 500 = 420$ bags

(e) $\frac{13-12}{0.5} = 2$
$\frac{1}{2}(95\%) + 50\% = 97.5\%$
$0.975 \times 500 = 487.5 \approx 488$ bags

(f) $\frac{13.5-12}{0.5} = 3$
$\frac{1}{2}(99.7\%) + 50\% = 99.85\%$
$0.9985 \times 500 = 499.25 \approx 499$ bags

53. (a) $\frac{15.25-17.25}{2} = -1$
16th percentile

(b) $\frac{21.25-17.25}{2} = 2$
97.5th percentile

(c) The 75th percentile corresponds to the third quartile.
weight = $17.25 + 0.675 \times 2 = 18.6$ lb

55. (a) $\frac{11-8.75}{1.1} \approx 2$
97.5th percentile

(b) $\frac{12-8.75}{1.1} \approx 3$

99.85th percentile

(c) The 25th percentile corresponds to the first quartile.
weight = $8.75 - 0.675 \times 1.1 \approx 8$ lb

E. The Honest- and Dishonest-Coin Principles

57. **(a)** $\mu = \frac{3600}{2} = 1800, \sigma = \frac{\sqrt{3600}}{2} = 30$

(b) $\frac{1770-1800}{30} = -1, \frac{1830-1800}{30} = 1$

68%

(c) $\frac{1}{2}(68\%) = 34\%$

(d) $\frac{1860-1800}{30} = 2$

$\frac{1}{2}(95\%) = 47.5\%$

$47.5\% - 34\% = 13.5\%$

59. $\mu = \frac{7056}{2} = 3528, \sigma = \frac{\sqrt{7056}}{2} = 42$

(a) $\frac{3486-3528}{42} = -1, \frac{3570-3528}{42} = 1$

68%

(b) $\frac{1}{2}(100\% - 68\%) = 16\%$

(c) $\frac{1}{2}(68\%) + 50\% = 84\%$

61. **(a)** $\mu = 600 \times 0.4 = 240,$

$\sigma = \sqrt{600 \times 0.4 \times (1-0.4)} = 12$

(b) $Q_1 \approx 240 - 0.675 \times 12 \approx 232$

$Q_3 \approx 240 + 0.675 \times 12 \approx 248$

(c) $\frac{216-240}{12} = -2, \frac{264-240}{12} = 2$

0.95

63. **(a)** $p = \frac{1}{6}$

$\mu = 180 \times \frac{1}{6} = 30, \sigma = \sqrt{180 \times \frac{1}{6} \times \left(1 - \frac{1}{6}\right)} = 5$

(b) $\frac{40-30}{5} = 2$

$\frac{1}{2}(1-0.95) = 0.025$

(c) $\frac{35-30}{5} = 1$

$\frac{1}{2}(0.68) = 0.34$

65. We assume that the defects are distributed normally so that the mean number of defects on a given day is $\mu = 1600p$ by the Dishonest-Coin principle. On the other hand, the mean number of defects will fall exactly between 117 and 139 defects. So, $\mu = 128$. Solving $1600p = 128$ for p gives $p = 128/1600 = 0.08$. That is, the probability that a randomly selected widget produced by this machine is defective is 8%.

JOGGING

67. **(a)** weight = $17.25 + 1.65 \times 2 = 20.55$ lb

(b) weight = $17.25 - 0.25 \times 2 = 16.75$ lb

69. **(a)** $\frac{17.75-17.25}{2} = 0.25$

60th percentile

(b) $\frac{16.2-17.25}{2} = -.52$

30th percentile

71. **(a)** score = 502 + (0.675)(113) = 578.275 ≈ 580 points

(b) score = 502 + (0.52)(113) = 560.76 ≈ 560 points

(b) $\frac{530-502}{113} \approx 0.248$. Thus, a test score of 530 is about ¼ of a standard deviation above the mean. This means a score of 530 is at the 60th percentile.

73. **(a)** The 90th percentile of the data is located at $\mu+1.28\times\sigma=65.2+1.28\times10=78$ points.

(b) The 70th percentile of the data is located at $\mu+0.52\times\sigma=65.2+0.52\times10=70.4$ points.

(c) The 30th percentile of the data is located at $\mu-0.52\times\sigma=65.2-0.52\times10=60$ points.

(d) The 5th percentile of the data is located at $\mu-1.65\times\sigma=65.2-1.65\times10=48.7$ points.

75. The mean number of tails tossed is $\mu=\frac{n}{2}$. There is a 16% chance Y is more than one standard deviation above the mean, $\frac{n}{2}+\sigma$.

$$\sigma=10$$

$$10=\frac{\sqrt{n}}{2}\quad\left(\text{using }\sigma=\frac{\sqrt{n}}{2}\right)$$

$$20=\sqrt{n}$$

$$n=400$$

Mini-Excursion 4

WALKING

A. Weighted Averages

1. Using the fact that 90/120 = 75% and 144/180 = 80%, Paul's score in the course is the weighted average $0.15\times77\%+0.15\times83\%+0.15\times91\%+0.1\times75\%+0.25\times87\%+0.2\times80\%=82.9\%$.

3. 100% - 7% - 22% - 24% - 23% - 19% = 5% are 19 years old. So, the average age at Thomas Jefferson HS is the weighted average $0.07\times14+0.22\times15+0.24\times16+0.23\times17+0.19\times18+0.05\times19=16.4$.

B. Expected Values

5. The expected value of this random variable is $E=\frac{1}{5}\times5+\frac{2}{5}\times10+\frac{2}{5}\times15=11$.

7. (a)

Outcome	$1	$5	$10	$20	$100
Probability	1/2	1/4	1/8	1/10	1/40

(b) $E=\frac{1}{2}\times\$1+\frac{1}{4}\times\$5+\frac{1}{8}\times\$10+\frac{1}{10}\times\$20+\frac{1}{40}\times\$100=\$7.50$

(c) $7.50

9. (a)

Outcome	0	1	2	3
Probability	1/8	3/8	3/8	1/8

(b) $E=\frac{1}{8}\times0+\frac{3}{8}\times1+\frac{3}{8}\times2+\frac{1}{8}\times3=1.5$ heads

11. (a)

Outcome (Profit)	$1 (red)	$(-1) (black)	$(-1) (green)
Probability	18/38	18/38	2/38

$E=\frac{18}{38}\times\$1+\frac{18}{38}\times\$(-1)+\frac{2}{38}\times\$(-1)=\$-\frac{1}{19}\approx\$-0.05$

(b) $E=\frac{18}{38}\times\$N+\frac{18}{38}\times\$(-N)+\frac{2}{38}\times\$(-N)=\$-\frac{1}{19}N\approx\$-0.05N$. So, for every $100 bet on red, you should expect to lose about $5 ($5.26 to be more precise). For every $1,000,000 bet on red, you should expect to lose about $52,631.

13. (a) $E=\frac{1}{6}\times\$1+\frac{1}{6}\times\$(-2)+\frac{1}{6}\times\$3+\frac{1}{6}\times\$(-4)+\frac{1}{6}\times\$5+\frac{1}{6}\times\$(-6)=\$-0.50$

(b) Pay $0.50 to play a game in which you roll a single die. If an odd number comes up, you have to pay the amount of your roll ($1, $3, or $5). If an even number (2, 4, or 6) comes up, you win the amount of your roll.

C. Miscellaneous

15.

Outcome (Benefit to Joe)	\$320	\$(-80)
Probability	24%	76%

$E = 0.24 \times \$320 + 0.76 \times \$(-80) = \$16$

Joe should take this risk since his expected payoff is greater than 0 (\$16). That is, if he made this transaction on thousands of plasma TVs, in the long run he could expect to save \$16 for each warranty he purchases.

17.

Payoff (to insurer)	\$*P*	\$(*P*-500)	\$(*P*-1,500)	\$(*P*-4,000)
Probability	50%	35%	12%	3%

In order to make an average profit of \$50 per policy, we solve $E = 0.5 \times \$P + 0.35 \times \$(P-500) + 0.12 \times \$(P-1{,}500) + 0.03 \times \$(P-4{,}000) = \$50$ for *P*. Doing this gives *P* = \$525. That is, the insurance company should charge \$525 per policy.

19. **(a)** There are three ways to select which of the three dice will not land as a 4. There are five numbers that this die can land on (1, 2, 3, 5, or 6). The multiplication rule tells us that there are $3 \times 5 = 15$ ways to select an outcome in which exactly two 4's are rolled. Since there are $6^3 = 216$ (again by the multiplication rule) possible outcomes in this random experiment, the probability of such an outcome is 15/216.

(b) There are three ways to select which of the three dice will land as a 4. There are five numbers that each other die can land on (1, 2, 3, 5, or 6). The multiplication rule tells us that there are $3 \times 5 \times 5 = 75$ ways to select an outcome in which exactly one 4 is rolled. Hence, the probability of such an outcome is 75/216.

(c) There are five numbers that each die can land on (1, 2, 3, 5, or 6). So, there are $5 \times 5 \times 5 = 125$ ways to select an outcome in which no 4 is rolled. So, the probability of not rolling any 4's is 125/216.

21.

Outcome (Winnings)	\$29,999,999	\$(-1)
Probability	$\frac{1}{{}_{47}C_5 \times {}_{27}C_1}$	$1 - \frac{1}{{}_{47}C_5 \times {}_{27}C_1}$

$$E = \frac{1}{{}_{47}C_5 \times {}_{27}C_1} \times \$29{,}999{,}999 + \left(1 - \frac{1}{{}_{47}C_5 \times {}_{27}C_1}\right) \times \$(-1) =$$

$$\frac{1}{41{,}416{,}353} \times \$29{,}999{,}999 + \frac{41{,}416{,}352}{41{,}416{,}353} \times \$(-1) \approx \$-0.28$$

That is, on each \$1 lottery ticket purchased, you should expect to lose about \$0.28. [In general, the question is slightly more complicated than this since it is not reasonable to expect never to need to split the jackpot. Also, in most lotteries of this sort there are other (lesser) prizes that can be won.]